AF677197

Wetland Restoration and Construction

A Technical Guide

To Angela,

Thanks for helping us restore wetlands.

Best wishes,

Thomas R. Bi...

By Thomas R. Biebighauser

Published by the Upper Susquehanna Coalition

Copyright 2011 © by Thomas R. Biebighauser

Edited by James Curatolo and Lucia Parr
Drawings by Amy Reges
Photographs by author unless otherwise noted
Manufactured in China

Biebighauser, Thomas R.
Wetland Restoration & Construction, A Technical Guide / Thomas R. Biebighauser.
p.cm
Includes bibliographical references and index.
ISBN: 978-0-9834558-0-6
Wetland restoration. 2. Construction techniques. I. Title

FSC
www.fsc.org
MIX
Packaging from responsible sources
FSC® C010953

OtterTail Art

Unless otherwise noted all graphical images within this book are the product and property of Amy Reges, OtterTail Art http://www.OtterTailArt.com. All rights reserved.

Partners in the development and distribution of this project include:

www.wetlandsandstreamrestoration.org

The Wetland Trust
www.thewetlandtrust.org

www.parcplace.org

www.u-s-c.org

CONTENTS

Acknowledgments ... 2

Introduction ... 3

Chapter 1 Why Should We Build Wetlands? ... 5
Chapter 2 Locating Drained Wetlands ... 10
Chapter 3 The Basics of Wetland Restoration ... 28
Chapter 4 The Construction of Surface Water Wetlands ... 40
Chapter 5 The Construction of Groundwater Wetlands ... 79
Chapter 6 Building Wetlands of Various Types ... 97
Chapter 7 Developing Wetlands Using Liners at Schools ... 107
Chapter 8 Establishing the Preferred Vegetation ... 121
Chapter 9 Designing Wetlands for Wildlife ... 129
Chapter 10 Obtaining Government Permits ... 137
Chapter 11 Renovation and Maintenance ... 143
Chapter 12 Stream Restoration ... 155
Chapter 13 Finding Funding ... 161

Glossary ... 163
Appendix ... 165
References ... 179
About the Author ... 182
Index ... 183

ACKNOWLEDGEMENTS

I want to thank my wife Dee for encouraging me to write this book, even though it took time from working on our farm. Jim Curatolo is deserving of all my gratitude for his support, his editing, and for making publication of this book possible. I am grateful to Amy Reges for creating the excellent detailed drawings. Lucia Parr's skills at editing are evident in every phrase, number, and comma. Melissa Yearick did an incredible job of designing the volume. I appreciate the patience of contractors Farris Osborne, Mike Osborne, and John Utterback in answering my many questions about construction. I want to express thanks to the many individuals who provided the photographs that add clarity to the text and illustrate the true beauty of wetlands. I thank David J. Bandrowski, Frank Bodkin, John Byrd, Mark Davis, Ernie Garcia, Dr. Purnima Govindarajulu, Brandt Gutermuth, Dr. Nina Hemphill, Richard Hunter, Jason Jennings, Jennifer Johnson, Shane Jones, Emily Lawson, Jeffrey Lewis, Sarma Liepins, Andy Maceyka, Melissa Mullins, George Morrison, Dr. Arthur P. Parola, Jr., Dan Taylor, Chris Yearick, and Elke Wind for their many fine suggestions and very welcome help. I especially am grateful to Elizabeth Bullock, Dr. Elizabeth Byers, Wes Mattox, Dr. Guenter Schuster, Alleyn Unversaw, and Melissa Yearick for their assistance in writing sections of this book. Lastly, I thank my supervisor District Ranger James David Manner for supporting an active wetland restoration program on the Daniel Boone National Forest.

Introduction

This emergent wetland was created on mined lands in Rowan County, Kentucky.

Wetlands are beautiful, fascinating places full of plant and animal life. Visit a wetland and you are apt to see wood ducks, shorebirds, frogs, and dragonflies. You might surprise a wading great blue heron seeking his lunch or spy a deer enjoying a drink. Wetlands can be found almost anywhere, near rivers and streams, on hillsides, and even atop mountains.

Many rare species of plants and animals depend on wetlands for their survival. Of the 256 listed as threatened or endangered by the U.S. Fish and Wildlife Service in 1991, 43 percent are wetland dependent.[1] The bog turtle, whooping crane, Indiana bat, and barrens top minnow are but a few examples of endangered wildlife that require the presence of wetlands.

Wetlands provide valuable nursery areas for fish such as salmon, largemouth bass, muskellunge, and northern pike. Wetlands filter runoff to maintain the clean water that aquatic organisms need, and provide abundant food important to their survival. Wetlands also reduce stream bank erosion by slowing runoff after rainfall events.

This ephemeral wetland was probably constructed by hand in the late 1800s to provide water for livestock. It is located in an old field on a mountaintop in West Virginia on the Monongahela National Forest.

Unfortunately, the landscape we see today contains far fewer wetlands than in earlier times. Experts report that less than one-half of the original wetlands in the contiguous 48 states remain.[2] Over 80 percent of these eco-systems have been destroyed in California, Indiana, Illinois, Iowa, Kentucky, Missouri, and Ohio.[3] Government estimates reveal that approximately 495,100 acres of freshwater emergent, shrub, and forested wetlands were eliminated in the United States from 1998 - 2004.[4]

Conversion of wetlands to other land uses continues across the nation under a permitting system administered by the U.S. Army Corps of Engineers.

Serious interest in wetland restoration began after 1937 with the formation of the nonprofit organization Ducks Unlimited, Inc. Its goal was simple: to increase waterfowl numbers by constructing wetland habitats throughout North America. Since that time, many government agencies, nonprofit organizations, and private land-owners have heard of the need and joined the cause by building and restoring wetlands.

This forested wetland was restored for only $1,400 on the Daniel Boone National Forest in Kentucky.

Currently, the majority of wetlands are constructed by developers required to do so to mitigate loss for legally-permitted wetland destruction. Relatively few wetlands are being built simply because it is a good thing to do. To reverse the downward trend in wetland acreage across North America, a groundswell of wetland restoration is needed by individuals who are being proactive, not because they have to, but simply because they want to help the environment.

It is important to note that there is no rule saying wetlands must be built only where they used to exist. Techniques are available to create naturally-functioning and -appearing wetlands in places where they may never have occurred.

This wetland is filled with trash near Piney River, Virginia.

Constructed wetlands many times fail simply because they do not hold water long enough for aquatic plants and soils to develop. With our current knowledge, there is no reason for this to happen. The methods described in this book will help you obtain the desired water depth and duration of flooding needed to sustain emergent, ephemeral, forested, shrub, or wet-meadow wetlands. Some may say that detailed instructions are not needed to make a wetland, but few have learned how to make a soufflé by simply looking at a picture.

This book has been written to help you build attractive and naturally-functioning wetlands the first time, every time. Practical techniques are described in detail so you can design and construct your own wetland without the need for professional assistance. The methods described are based on experiences gained by professional engineers, land-owners, and volunteers in building and repairing thousands of wetlands across North America.

Be assured that you can build a wetland. The main qualifications needed are persistence, a conviction to help the environment, and a viable space. Do not let others sway you into believing that you have to be a professional engineer or a wetland ecologist to succeed. Many of the wetlands pictured in this book were built by individuals with little, if any, experience, but who shared a love for the environment, and were willing to jump in and get both feet wet to get started.

Mike Hayslett (shown here) built this emergent wetland used by the mole salamander (Ambystoma talipodium) for breeding near Piney River, Virginia.

Susan Guynn (shown here) built this wetland with waters clouded by deer on the Clemson Experimental Forest in South Carolina.

Science teacher Beverley McDavid (shown here) has built four wetlands at schools in eastern Kentucky for environmental education.

This is one of hundreds of wetlands built by Melissa and Chris Yearick (shown here) in New York.

Jim Ano built this small ephemeral wetland by hand at his home in Cincinnati, Ohio. (Jim Ano photo)

Forest Service Wildlife Biologist Jay Martin and Wildlife Technician Cheryl Tanner (shown here) found wood frog eggs in this wetland they built on a mountaintop in West Virginia on the Monongahela National Forest.

Mark Biser (shown here) shows one of the many wetlands he has built with a small excavator on his ranch in Washington State. (Mark Biser photo)

Teacher Tom Smith (shown here) built this wetland now used by wood frogs on mined land near Barbourville High School.

Dwayne Pergrem (shown here) and his father built this wetland in Bath County, Kentucky in 1955.

Dr. Art Parola and Danny Peake (shown here) examine a pool in a stream they restored along the Dix River in Lincoln County, Kentucky.

People are discovering that restoring wetlands does not have to be complicated or expensive. Landowners and public land managers with little experience who have a strong interest in helping the environment are building wetlands that look natural, attract a variety of plants and animals, and require little, if any, maintenance.

Chapter 1

Why Should We Build Wetlands?

Historically, it has not been possible to protect all wetlands from being drained legally, as well as illegally, across the nation. However, these millions of acres of changed land provide a tremendous opportunity for restoration. Indeed, the large tracts of public lands acquired over the past 100 years offer landscape-scale opportunities for ecosystem restoration.

It is possible to build wetlands that look and function like natural wetlands. Beaver have been building wetlands for thousands of years, and with our understanding of soil texture, groundwater, and construction techniques, we can now build wetlands even in places where beaver cannot.

Wetlands can be designed to treat waste, reduce flooding, replace groundwater, and provide nature enthusiasts with opportunities to see wildlife. We can now build wetlands to provide habitats for rare species, as well as the more commonly hunted and photographed wildlife.

While planning a wetland project, be careful that others do not lead one into an abyss of endless analysis which can only lead to inaction. One will meet many a naysayer along the path of wetland restoration that excels at throwing stones, providing no direction or encouragement. The historical record shows that millions of acres of wetlands were expertly drained by individuals with little or no experience. So why do some espouse that an advanced degree is required in order to construct one?

Some organizations who are currently funding wetland projects emphasize huge under takings involving a litany of partners that affect thousands of acres. While these efforts are very worthy, their scale can frighten the individual considering a wetland project for the first time. Remember that it is okay to restore a small wetland; in fact, it is preferable to start by building one that can be completed in one day. Building a small wetland involves the same steps as constructing a large wetland, except that it is less complex, and requires a fraction of the cost.

Small wetlands provide large values to plants and animals. Historically, the very small wetlands were among the first to be drained across the landscape.[5] Even today, most wetlands drained and filled are small. From 1998 to 2004, 52 percent of all freshwater wetlands lost in the United States were less than one acre in size.[6]

Wetlands are areas containing shallow water on or near the surface of the ground during all or part of the year, plants that are adapted to living in water or on wet ground, and soil that is saturated or poorly drained for a portion of the year. As much as possible, the various types of wetlands described in this book are named using the *Classification of Wetlands and Deepwater Habitats of the United States* published by the U.S. Fish and Wildlife Service.[7] In addition, areas containing water seasonally, known as ephemeral wetlands, not described by Cowardin, *et. al.,* are included in the definitions of wetlands.

Emergent Wetland

Water may be present all year. Soils are saturated and are generally gray or black in color. Common plants include water lily, duckweed, pondweed, pickerel weed, coontail, cattail, sedge, bulrush, and rose mallow. Common wildlife species include bald eagles, osprey, ducks, geese, herons, shorebirds, muskrats, mink, beaver, otters, bats, turtles, frogs, and salamanders.

This emergent wetland was restored in 2008 in an open field on Terry Back's farm in Bath County, Kentucky.

This natural emergent wetland surrounded by shrub wetland is near the Housatonic River in Massachusetts.

Ephemeral Wetland

Water is present during part of the year, generally in the winter and spring, drying in fall or under drought conditions. Soils are saturated in some, but not in others, and are often firm underfoot. They may be gray or brown in color. Common plants include sedge, bulrush, aquatic grasses, moss, wild iris, and rose mallow. Wildlife includes frogs, toads, salamanders, snakes, turtles, bats, fairy shrimp, and dragonflies.

This natural ephemeral wetland, shown in winter, is in Rowan County, Kentucky.

This wet meadow wetland which is dominated by sweet flag is located in Rockcastle County, Kentucky.

Wet-Meadow Wetland

Standing water can be found after flooding in small depressions. Soils are saturated and soft, often gray or black in color. Common plants include sedge, bulrush, rose mallow, monkey flower, cardinal flower, and aquatic grasses. Wildlife includes sparrows, finches, sand hill cranes, mice, voles, turtles, and snakes.

This restored wet-meadow wetland is located in Rowan County, Kentucky.

Shrub Wetland

Standing water can be found in small depressions. Soils are seasonally saturated, generally gray or black in color. Common plants include alder, willow, dogwood, buttonbush, sedge, and bulrush. Wildlife species include American woodcock, flycatchers, warblers, rabbits, mice, voles, and snakes.

Here speckled alder and buttonbush are growing in a shrub wetland in Rockcastle County, Kentucky.

This shrub wetland is fertile with willow and alder in Bath County, Kentucky.

Forested Wetland

Often flooded under high water conditions, a forested wetland contains seasonal and permanent pools of water. Soils are generally firm underfoot with some areas being saturated. Soil colors may be gray, black, or brown. Common plants include red maple, cypress, sweet gum, swamp white oak, box elder, river birch, sedge, bulrush, moss, and lady slipper. Wildlife species include warblers, flycatchers, woodpeckers, raccoon, and waterfowl.

This forested wetland, containing pit and mound topography, is located on the Monongahela National Forest in West Virginia.

This forested wetland is on the Clemson Experimental Forest in South Carolina.

This bog wetland is located at Cranberry Glades on the Monongahela National Forest in West Virginia.

Peat Lands

Standing water generally is not found, but may occur in small pockets. Soils are saturated and spongy. Common plants include sphagnum moss, leather leaf, Labrador tea, sedge, bulrush, pitcher plant, sundew, and orchids. Wildlife may include voles, sparrows, and finches.

These are scattered trees and shrubs at Cranberry Glades bog.

Wetlands are often made up of a variety of types. For example, a beaver pond generally contains an emergent wetland in the center, wet-meadow next to the emergent wetland, with shrub and forested wetlands on higher ground along the edge. Wetlands can be thought of as a continuum of related communities, with one type grading into another.

This demonstrates a transition between shrub and bog wetland.

Individuals across North America are building wetlands to accomplish functions such as flood control, water purification, groundwater recharge, and wildlife habitat improvement. They are restoring a variety of wetland types to help both common and rare species of plants and animals. An added bonus is that the ecosystems they construct are enhancing the beauty of the communities where they live.

Locating Drained Wetlands

Chapter 2

Restoring a wetland in its original location is one way of ensuring one's efforts will thrive. When rebuilding on a historic site, the odds of success in returning life can be increased by the seeds and crustaceans that may lie dormant in the soil. When searching for locations to build wetlands, it is generally accepted that priority be placed on finding sites where wetlands used to exist. Unfortunately, it is often difficult to prove that a location was once a wetland. Perhaps the only way to be sure is to talk to the person who drained it, or to a landowner who remembers its existence. Because the majority of wetlands were drained in the 1800s, one can knock on a lot of doors, and never find someone old enough to remember where they were once found.

In working to identify wetland restoration sites, some individuals look for basins that are growing aquatic plants, contain saturated soils, and perhaps, an obvious ditch running down the middle. It is important to remember that these areas represent the partially drained, failed attempts to convert a wetland to another use, and that the wetlands that were successfully converted to other uses do not look like wetlands anymore.

A wetland that was successfully drained no longer looks like a wetland. Gone are the aquatic plants, standing water, groundwater, and gray-colored soils. Interview those who have drained wetlands and one will discover that, once started, they usually finished the job. At first glance, drained wetlands look like any other agricultural field, woodlot, or housing development. Some expect to find drained wetlands by comparing old with new aerial photographs. While the technique can help in some situations, it has serious limitations because aerial photography did not come into widespread use until the 1930s, years after a majority of wetlands had been drained across North America. Those who insist on using photographs as the sole means of identifying restoration sites will miss all the wetlands drained before aerial photography.

Survey a natural wetland and one will find that its bottom contains a gradual slope. Wade down the slope in the wetland and one will find where its waters are backed up by a wide, natural dam with gradual slopes. People who drained wetlands learned to recognize where these natural dams were located, so they could build ditches and bury drain lines to release surface and groundwater.

Drained wetlands can be quite small, often only one-quarter acre in size. The reason people took the trouble to drain these tiny sites can be understood by considering that early agricultural fields, particularly those in mountainous areas, were small, and there was much to be gained by farming any available level piece of ground. Small drained wetlands can be difficult to find, and unfortunately, when trees have overgrown them, they are easily overlooked.

Do not expect a drained wetland to look like a wetland anymore. In general, such a high-quality job was done that all standing water, saturated soils, and aquatic plants have been eliminated from the site.

Cornfields in mountainous areas were created by moving creeks, along with draining and filling wetlands, a common practice.

Fortunately, there are clues that betray the presence of drained wetlands. These signs are summarized after interviewing individuals who worked to drain wetlands all their lives, by examining known drained wetlands across North America, and by researching drainage literature. With a little practice, one can spot the ghosts of drained wetlands while driving 70 miles per hour.

Small Wetlands

The presence of a small wetland often shows where a much larger and different type of wetland was once located. Small wetlands often develop in drainage ditches and along roadsides. These saturated patches of ground may also indicate where a buried drain line that is needed to eliminate a larger wetland is blocked. Water pressure in the drain line can build uphill from where the blockage is located and force water to the surface through gaps between the pipes. In 1903, drainage author Henry French described a method for finding obstructions in buried drain lines by punching a hole in the soil along the course of the drain. Water would burst up from the hole like a spring at the point of greatest pressure just uphill from the blockage, while there would be no upward flow of water below the blockage.[8]

Modified Wetlands

Examination of existing wetlands can often reveal signs of human activities affecting the hydrology, plants, and animals in that environment. In many cases, efforts to drain a wetland were only partially successful; however, these actions continue to affect the wetland years later.

Mike Hayslett has been monitoring amphibian reproduction in a series of natural ridge-top wetlands in Virginia for almost 20 years. One ephemeral wetland in particular had problems. Frogs, toads, and salamanders laid their eggs in its waters each spring, yet few emerged as adults. The difficulty was that the wetland did not contain water long enough into the summer. Mike and the author walked around the wetland in the spring of 2007, and found a shallow ditch leading from the wetland. Now grown up to trees and only ten inches deep, the feature was difficult to see. Apparently hand-dug years ago, the ditch was controlling the depth of water in the wetland. By working with the landowner to fill the ditch, the increase in dcpth should improve breeding conditions for herps.

Fortunately, steps can be taken to return a more natural condition to an altered wetland that will benefit plant and animal life. The following signs indicate the site of a wetland that has been modified by previous human activities:

Runoff is prevented from reaching a wetland by the presence of roads, roadside ditches, railroad grades, or drainage ditches upstream. Even the tracks of Off Highway Vehicles (OHVs) can divert small streams that have long supplied a wetland, changing entire plant and animal communities in that wetland. Vehicle tracks can also concentrate the normal sheet flow of water moving overland to a wetland into deep, narrow channels that carry soils into the wetland.

- Finding wetlands present on both sides of a road indicates the road was built through a wetland. Roads can change how groundwater moves in a wetland. Groundwater moving through peat lands, forested, and shrub-scrub wetlands can be blocked by the compacted soils in a road or trail. Roads act like dams to raise the water table upslope, making conditions too wet for trees such as black spruce, tamarack, and white cedar. Conversely, downstream areas may become drier and grow unnaturally dense thickets of trees and shrubs when waters are blocked.

- Attempts to drain a wetland are usually visible, which explains why waters are often found, leaving a wetland via constructed ditches and buried drain lines. To the casual observer, some ditches can appear to be natural streams; however, their straight course betrays a human design.

- Finding a dam with steep slopes can indicate where a natural wetland was changed to create a ranch or farm pond. Discovering a wetland that is long and narrow with straight edges is evidence that the wetland was formed in a drainage ditch.

- Ruts from vehicles in a wetland can form unnatural zones of compacted soils. In southern areas, ruts can form narrow strips of permanent waters that serve as refuge for predatory fish, which in turn decimate populations of pond-breeding salamanders, frogs, and toads. Non-native carp can also survive in these ruts, which can reduce plant abundance and diversity in the entire wetland. In Alaska, salmon fry can become stranded in ruts as they dry out.[9]

Individuals have damaged this natural wetland by driving vehicles through it repeatedly.

- Ruts can adversely affect water flowing into, within, and out from wetlands. Water running down vehicle tracks may indicate where a stream has become a travel route, or where the stream changed course and now follows an unplanned route.

- Finding crushed aquatic plants and large areas of exposed soils often shows where vehicles are causing damage to wetlands. Several passes of a motor vehicle can kill colonies of aquatic plants, and the invertebrates that use these plants for protection and food. OHV traffic can compact soils, thereby decreasing plant diversity. Soils found in partially dried wetlands are easily compacted by the rubber tires found on most vehicles, and these compacted soils are generally unsuitable for plant growth. Observing scattered clumps of aquatic plants growing in puddles with ruts and tire tracks indicates where wetlands are being damaged by this vehicle activity.

- Finding an abundance of non-native plants may indicate where vehicles or horses are transporting unwanted plants into the wetland. The deep treads system found on most vehicle tires hold mud, and consequently seeds, from previously traveled routes. If not washed before transporting on the highway, invasive plants can be spread over each subsequently traveled landscape.

- Muddy waters in a wetland may indicate the presence of carp, which will eliminate aquatic plants over time. Muddy waters may also be caused by vehicle traffic. Waters high in sediment can impact plant, amphibian, and invertebrate populations.

- Absence of animal species one would normally expect to find in a wetland, such as dragonfly larvae and frog tadpoles, may indicate a problem with water quality in the wetland.

One generally greatly underestimates the extent of actions taken to drain a piece of land. For each ditch or buried drain line outlet found, there are probably ten not found. When designing a wetland project, it is wise to assume that generations of individuals toiled to remove excess water from the property and that, like a puzzle, it will take patience and perseverance to piece the wetland together again.

It should come as no surprise that the majority of sites individuals select for building wetlands were changed historically by drainage activities. In most situations, the somewhat level areas of ground people choose for projects would already be wetlands if the moved creeks, ditches, and buried drainage structures were not still functioning as intended by previous owners.

Buried Drain Lines

Discovering clay tiles or plastic pipe strongly indicates that an area was a wetland. Drain lines are generally buried 30 inches or more in the ground to remove both surface and groundwater from the soil. They often start uphill from a wetland, pass through the wetland, and carry water downhill to an outlet, which can be a ditch, stream, or river. Drain lines effectively lower the water table, remove standing water from large and small areas, and can remain effective for hundreds of years with little or no maintenance.

Sometimes, one can get lucky and actually find a drain line outlet by walking the creek bank. However, this method is seldom successful. Creek banks are usually overgrown with shrubs, and trees, and are so steep and muddy that they can be difficult to examine. An exposed pipe with water flowing from it is clearly an outlet. But many outlets may only be visible as wet spots in a creek bank because they are buried. Outlets will continue to function, even if they are under water and covered with soil.

Locating an outlet indicates that extensive measures were taken to drain wetlands uphill from the site. Large drainage systems are typically emptied into a single outlet in order to reduce maintenance costs, and to prolong the life of a system.[10]

Natural wetland

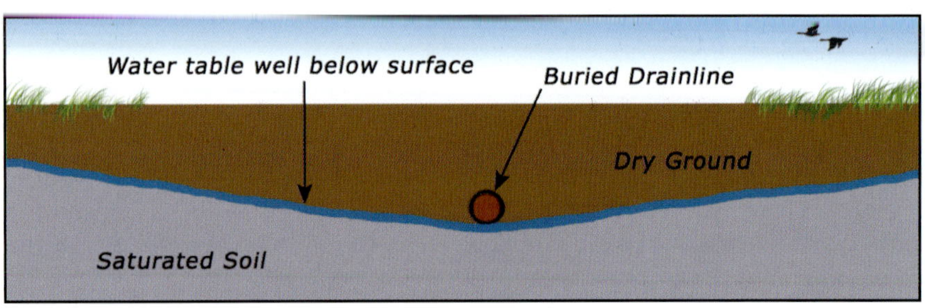

Same wetland eliminated by burying drain lines, filling, and leveling

This outlet from a four-inch diameter system of clay tiles buried in the 1800s to drain wetlands is located along Stonecoal Creek in Rowan County, Kentucky.

This outlet from a four-inch diameter plastic drainage system used to dry wetland is in Carter County, Kentucky.

Even after a field has grown up to trees, buried drain lines will continue working. Henry French wrote "The behavior of roots is, however, very capricious in this matter; for, while occasional instances occur of drains being obstructed by them, it is a very common thing for drains to operate perfectly for indefinite periods, where they run through forests and orchards for long distances."[11]

> *"On the whole, after wide observation, I am prepared to assert anew, that tile-drainage will endure forever if the work be properly done."*[12]
> *Henry F. French, Drainage Engineer.*

Buried clay drain tiles continue to function in this young forest, once a wetland cleared and drained for farming in Louisville, Kentucky.

Buried drain lines can be difficult to detect without excavation. However, experiments have been conducted to locate buried drain lines by using ground radar with limited success. The radar units, which are calibrated to detect possible underground channels, are pulled on trailers over the ground. This research focused on helping farmers locate buried lines in need of repair, but could also help individuals find buried lines responsible for wetland failure.[13]

Vertical Pipes

A vertical, short section of pipe with holes in it above the ground generally marks the presence of a drained wetland. Known as a Hickenbottom, or surface inlet, these are used to rapidly drain surface water from a wetland into a system of drain lines that carry water underground to a ditch or creek.

Finding the entrance of a vertical pipe in the ground can also indicate the presence of an agricultural drainage well. Wetlands that occurred in depressions where it was not possible to drain the waters downhill into a creek were often eliminated by digging a vertical hole in the ground, whose depth could range from six feet to over hundreds of feet deep to empty waters into a permeable layer of sand, rock, or gravel. Expect the sides of the drilled hole to be protected from collapse by pipe casing. Vertical wells can provide a permanent path for waters to leave the wetland. Drain lines are generally buried throughout the bottom of the wetland to carry waters down into the drainage well. The entrance of a well is usually protected by a grate; these wells will continue to function even when the grates are covered with stone and soil.

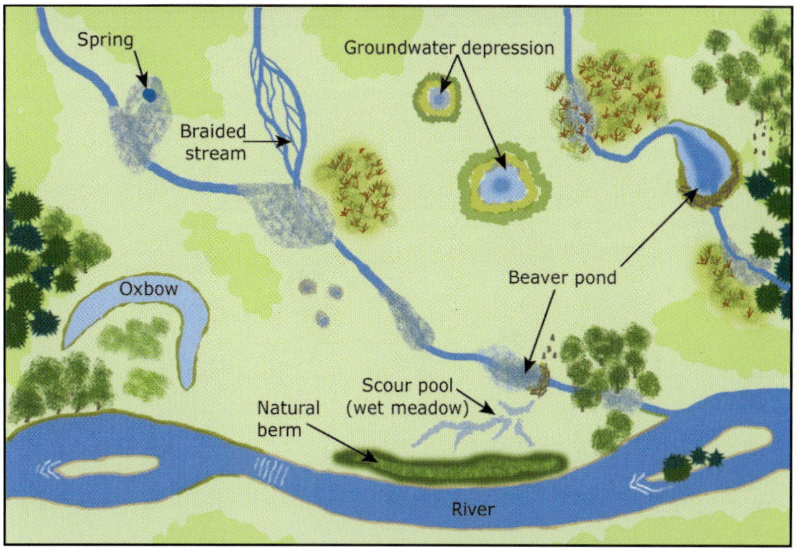

Natural Landscape (above) and Modified Landscape (below)

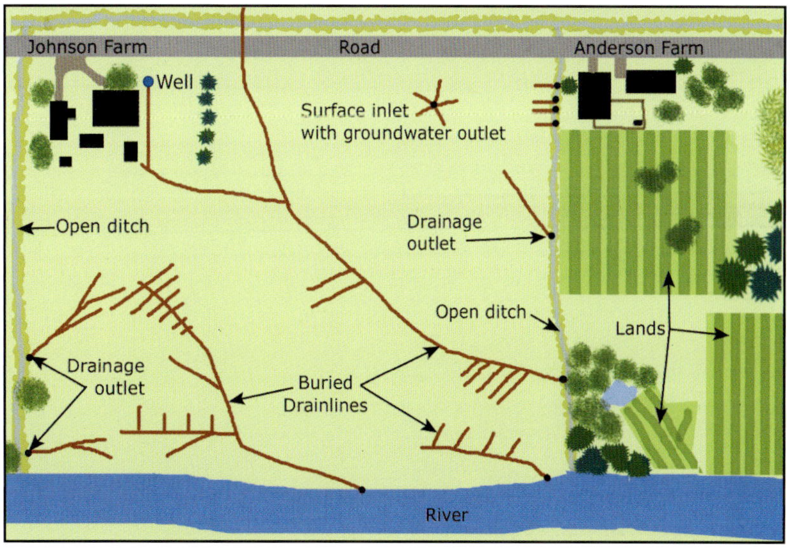

This surface inlet installed in a ditch diverts runoff into buried drain lines for drying adjoining farmland near Pine Swamp, Pennsylvania.

The horizontal hole, 18 inches below the surface, shows the still functioning subsurface drainage system, probably created using a mole plow. (Cynthia Sandeno photo)

The outlet from the same subsurface drainage system carries a steady stream of water into the relocated creek, now a drainage ditch. The drainage system was probably created using a mole plow years ago. (Cynthia Sandeno photo)

Vertical Holes

Observing holes in the ground not created by animals often indicates the presence of structures buried to drain

This vertical hole connects with a subsurface, horizontal drainage ditch probably created using a mole plow. A series of vertical holes were spaced along a straight line between a dry depression, indicating a historic wetland, and a moved creek on the Monongahela National Forest in West Virginia. (Cynthia Sandeno photo)

wetlands. These holes on the surface can be six inches or larger in diameter, and may either be isolated or have several spaced along a straight line. Vertical holes may form above loose joints between clay tiles, over rotted sections of wood drains, rock drains, or above horizontal holes created with a mole plow. These vertical holes, also known as blowouts, can be almost any depth. In some cases, the structures used to drain the wetland can be seen in the bottom of these holes. During wet times of the year, one may see and hear water flowing in the bottom of these holes.

Buried Rock

Finding a patch or buried column of fieldstone, crushed gravel, or creek rock can show where a wetland was located. The rock can indicate the entrance of an inlet created for draining water from the deepest part of the wetland straight down into the ground, or down slope to a creek, river, or ditch some distance away. Coils of plastic drain lines were often buried beneath the rock to collect waters in order to dry an area for cultivation.

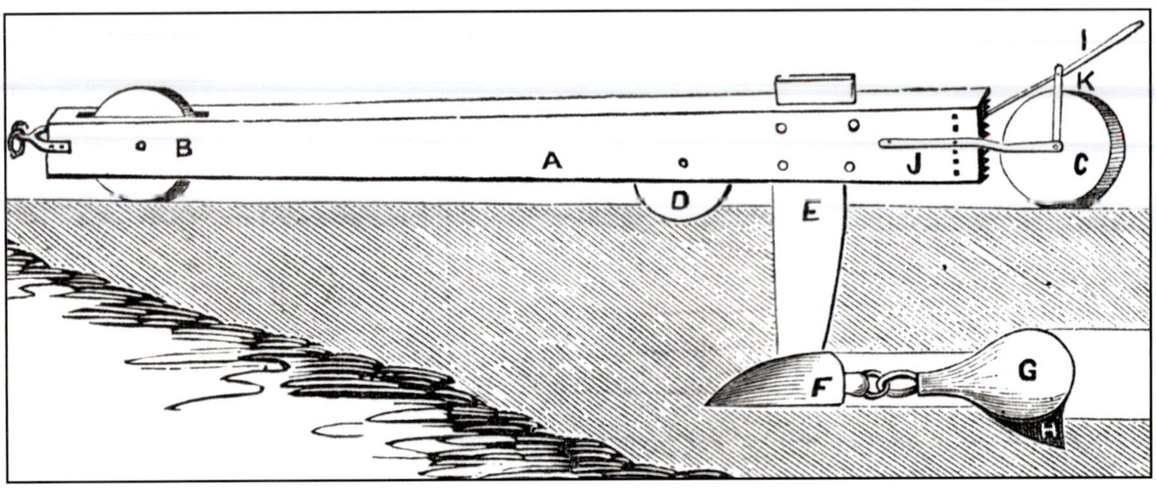

This is a mole plow being used for drainage, from John H. Klippart's 1861 book The Principles and Practice of Land Drainage[14]

(left) This rock drain was uncovered during a wetland restoration project on the Monongahela National Forest in West Virginia. (Cynthia Sandeno photo)
(right) A line of rocks mark the top of a drain used to dry a wetland in Victoria, British Columbia on the Haliburton Organic Farm. Two drains, approximately 12 inches wide by 18 inches deep and filled with thousands of smooth rocks were uncovered while restoring wetlands on the site. (Andrea Paetow photo)

Wells

An open or covered well may indicate the presence of a drainage system used to eliminate wetlands. Deep wells were dug to serve as sediment traps for tile systems, and as places to combine waters flowing from smaller lines into a larger line. Wells can be of various sizes, and many were constructed large enough to be entered for cleaning.[15]

Pumping Stations

The presence of a pumping station provides strong evidence that a wetland occurred on the site. Pumps were used to remove waters where no other outlet was available that could use gravity for the task. The pump was placed in a sump that collected water from a system of ditches and buried drain lines. It then moved the water to a ditch or stream further away. An electric line and power pole are often located near the pump and its collection basin. Many pumps were installed by a Drainage District years ago, and are now maintained by the landowner who receives the greatest benefit from the system.

A pumping station in the center of the field show the area was once largely wetland.

Gaps Along the Riverbank

Examining how side drainages enter the main channel of a river or stream can aid in locating former wetlands. If a smaller stream enters the main channel at a right angle, and its channel is straight, it probably represents a constructed outlet used to drain wetlands. These outlets were designed to remove standing water trapped behind the natural rise or dam along the creek bank. In comparison, an unaltered creek's entrance should be higher, contain braided shallow channels with gradually sloped banks, and look more like a delta.

This drainage ditch enters Beaver Creek at a right angle in Menifee County, Kentucky. Fields created on either side were farmed over 40 years ago.

Crayfish Burrows

The presence of crayfish burrows, active or inactive, can indicate the location of a drained wetland. Their burrows tell us that the site once had, or still has, a high water table and areas of standing water. Crayfish found burrowing in agricultural and other fields, and the ditches associated with the draining of these fields, are quite possibly the descendents of those present prior to drainage. "Finding a field with a ton of crayfish tells me the area used to be wetland," says crayfish expert Dr. Guenter Schuster.[16]

Drainage practices harm crayfish by eliminating areas of standing water, lowering the water table, and reducing the presence of moist soils. Crayfish must keep their gills moist for respiration, and generally accomplish this by staying in water. Those species considered primary and secondary burrowers can survive outside of water under conditions of high humidity, explaining why they are sometimes found on the surface during wet spring and fall nights. Under drought or freezing conditions crayfish must retreat to burrows, resting in waters contained within the burrows. Because drainage activities lower the water table, crayfish must dig ever deeper to survive in destroyed wetlands. Schuster believes that soils dried by drainage activities may become unsuitable for crayfish because it becomes too difficult for them to dig burrows.[17]

After a wetland is drained, farming operations that continue over the surface may maintain conditions that allow crayfish to continue using a site. Disking and plowing may keep soils moist and loose so that crayfish can continue to dig burrows necessary for reaching groundwater. A dense growth of tree roots appears to make locations unsuitable for at least some species of crayfish, and farming also keeps trees from taking over the site.

Young crayfish, which are not as strong as adults, probably have a more difficult time surviving in drained wetlands. They are less likely to displace adults already using burrows, and cannot dig new burrows in hard, dry soils. One may think crayfish can easily survive a wetland drainage project by dispersing into other areas, but movement by primary burrowers is limited to walking overland, and on a drained landscape other habitats are generally not available. Traveling away from an existing network of burrows can result in death by drying or by predators.[18]

Bluegrass crayfish (Cambarus batchi), is considered to be a secondary burrower, and whose presence in agricultural fields indicates historic wetlands. (Guenter A. Schuster photo)

This chimney surrounds a burrow made by the valley flame crayfish (Cambarus deweesae). (Guenter A. Schuster photo)

Ditches

The primary reason for digging a ditch is to attract water from wet ground so that it can dry. Ditches are highly effective for removing both surface and groundwater. A shallow ditch that is difficult to see can divert runoff from a hillside, as well as absorb standing water. Deep ditches can also lower the water table and provide an outlet for waters to be removed from surrounding buried drain lines. Generally, a combination of shallow, barely discernable ditches and more obvious deep ditches are used to drain wetlands.

Ditches are dug to drain wetlands. The presence of a ditch indicates the surrounding ground used to be wet, very wet.

People dug ditches because there was no other choice. It is generally not possible to raise crops on a wetland without ditching. Digging a ditch by hand to drain a wetland is hard work. It is extremely difficult to shovel the wet, heavy soils while standing in water, and it is safe to say that these projects were not undertaken unless water was a real problem.

Ditches really work. A shallow, six-inch deep ditch along the base of a hill can dry an entire field. Deeper ditches can remove standing water and lower a high water table in only a couple of days. Finding a ditch is clear proof that the site used to be wet, and in all probability, was a wetland.

The diversion ditch dug at the base of the hill to drain a wetland continues to function long after the field has grown up to trees.

The presence of a ditch provides clear indication that land leveling and reshaping also took place. One does not dig a ditch unless all of the waters in the surrounding area will flow into the ditch. Entire fields were routinely reshaped so that water would runoff into ditches. Slopes were placed on flat lands and depressions filled so that rain and snowmelt would flow into ditches.

Digging a ditch produces a lot of soil, and the most logical place to dispose of this soil is to use it to fill in low places

Natural land

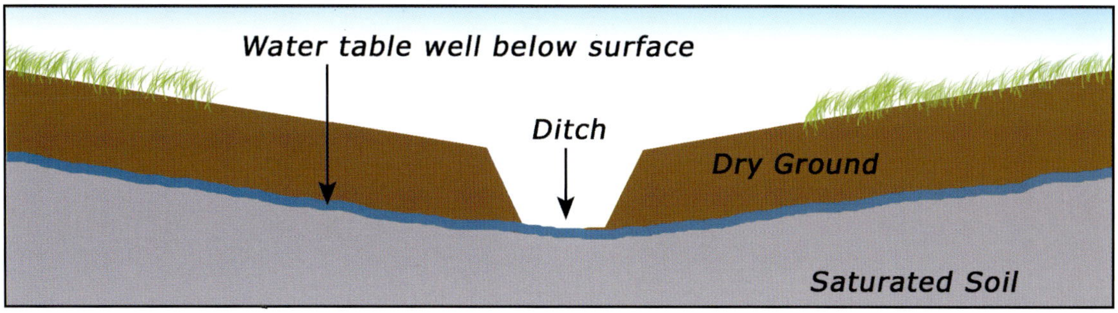

Same wetland eliminated by drainage ditch and contouring

where water can stand. Raising low places with excess soil is standard practice. This is the reason it is not enough to simply fill ditches when restoring a wetland. The land should also be reshaped to return natural basins that had been filled and sloped.

The depth, pattern, and distance between ditches can indicate how wet the land used to be. Shallow ditches show where there was a problem with water standing on the surface. Deeper ditches demonstrate historic problems with both surface and groundwater. Closely spaced ditches mark where the ground was very wet.

The lack of ditches does not mean that a site was never a wetland. Waters were also often removed through an underground system of buried drain lines. Finding a pipe or outlet in a ditch or along a stream shows that a network of buried drain lines had been installed uphill. The outlet for a buried drainage system can be hard to find unless its location is known by the landowner or someone else. In many cases, the actual outlet will become buried, and this underground system can continue to function indefinitely.

Clay pipe in a ditch performs as the outlet for a system of buried clay tiles draining wetland at Sweet Briar College in Virginia.

The presence of long, deep ditches indicates that wetlands have been drained over a large area. These ditches are basically straightened streams designed to carry large amounts of water. The deep ditches often cross lands owned by a number of individuals, and can appear quite natural, being bordered by large trees and containing aquatic vegetation. Some are even labeled as streams on topographic maps. They were generally not dug to drain wetlands directly, but provided the means by which a number of smaller wetlands on the surrounding landscape could be eliminated. The deep ditches provided the outlet essential for pulling waters from a system of buried drain lines located upslope that were used to remove surface and groundwater from wetlands.

Deep ditches were often the first and most important action that enabled landowners to remove water from their wetlands without fear of flooding their neighbors downstream. In mountainous areas, these ditches may be less than one-quarter mile long; on more level terrain, they can run for miles.

This deep, long ditch was historically constructed to serve as an outlet for multiple, adjoining buried drainage systems. Its presence shows that large acreages of wetlands have been eliminated across the landscape.

The presence of shallow ditches also indicates a site where wetlands were drained. Shallow ditches prevent waters from pooling that can form wetlands. A field or woodlot containing a series of shallow, parallel ditches was probably once a wetland. Ditches can be difficult to detect after an area has grown up to shrubs and trees. However, they only need to be a few inches deep to divert water, and some carry water only after a heavy rain. Ditches were often dug along the base of a hill to keep runoff from maintaining wetlands on the more level ground below. Ditches were also used on slopes to divert waters from wet-meadow and shrub wetlands.

Finding an open ditch that ends before reaching a creek often indicates the presence of a short buried drain line leading from the ditch. The buried line carries water from the ditch and drained wetland downhill to a creek and acts as a culvert to provide equipment access to fields.

Warning, do not be misguided by calling a ditch a wetland. It is unfortunate when individuals who think they are doing a good deed mistakenly protect leftover wetlands found in drainage ditches. Ditches often contain aquatic plants, gray-colored soils, and small pools of water. They can be narrow or wide, have steep or gradual sides, and traverse steep or level ground. Clues to the presence of ditches include a straight line appearance and narrow transition between wet and dry ground.

"Lands" Plowing

Finding a series of shallow ditches that parallel each other from 30 to 100 feet apart shows that an area was once so wet that directional plowing was used to create ridges for planting. The peak of these ridges can be from six inches to three feet higher than the bordering ditches. These ditches may or may not empty into a deeper ditch or creek. Farmers call these ditches that do not drain "dead furrows". This practice of plowing, known as "lands", was used to raise crops on areas of wetland, and generally indicates that more effective buried drain lines are not present. Narrow lines of bulrushes, sedges, and willows will often grow in the shallow ditches between the slightly higher ridges of soil. These "green streaks" of aquatic plants may parallel each other over the entire length of a field, often disappearing when the field is plowed or cut for hay.

The "lands" pattern is more difficult to detect when an area is no longer farmed. Trees and shrubs soon grow in the bottom of these ditches, and they can look like grown up fencerows. These rows of trees may even be present in the ditches when the area is being farmed, generally indicating that an individual with less concern for the field is now tending the farm. Keeping trees from growing in the bottom of the ditches along the base of "lands" requires more work than simply cutting hay or harvesting corn on a tractor. The farmer has to pay close attention to the weather and mow these ditches when it is dry enough not to find his heavy equipment stuck in the mud. In wet years, it is necessary to cut the trees by hand, for if trees are allowed to grow in the ditches, they will soon reduce crop production by shading the field.

This historic wetland is being farmed in "lands", evidenced by long ridges bordered by shallow, parallel ditches.

Springs

Look for springs along the base of hills. They are often associated with old homes and barns, where their waters were critical for drinking and for livestock. Waters flowing from springs generally form extensive wetlands and streams. If they do not, the site was probably modified for drainage.

Spring-fed wetlands were drained on this farm in Lawrence County, Ohio by placing waters underground in pipes into the cement watering trough shown below. Overflow from the trough was then piped underground to the creek, which was moved. The original source of the spring is marked by cattle-trampled soils behind the tree.

This cement watering trough on a hillside indicates the presence of historic wetland on the Larry and Sharon Kearns farm in West Virginia.

John D. Smith (right) stands where a spring once emerged that produced a wetland on his farm in Menifee County, Kentucky. He placed waters from the spring underground in clay drain tiles, and remembers the creek flowing below the large white oak behind him.

Finding a spring that does not saturate the surrounding ground, or whose overflow does not flow as a creek indicates a drained wetland. Springs are responsible for maintaining many wetlands; farmers recognized this fact and took measures to carry runoff from springs beneath the ground directly to streams, thereby draining the wetlands supplied by them.

Watering troughs and small ponds located on the sides of hills indicate where springs have been developed to supply water for humans and livestock. Pipes are generally buried underground to direct waters into troughs, and to drain excess waters downhill. Springhouses show where waters flowing to the surface were harnessed to cool perishables for homesteads, and where overflow was also directed underground to nearby streams and ditches. The presence of water troughs and springhouses indicates a continuous flow of water that once supplied wetlands.

A springhouse near Watkins Glen, New York shows where waters once supplying wetlands are now being collected and directed beneath the road in pipes to the creek.

Unplanted Areas in Fields

One clue to the presence of a drained wetland is spotting an area in a field that is not planted or mowed. Trees and shrubs growing in these locations often represent the deepest portions of a much larger wetland drained years ago. One may find rocks, logs, and brush, and even trash

A clump of trees growing in a soybean field marks the deepest part of a wetland now drained by the ditch in the foreground in Chester County, Pennsylvania.

buried in these areas, as they were considered to be nonproductive. In a field that is growing up to shrubs and trees, these historically unplanted places appear as clumps of much larger trees.

This lone, old growth pin oak betrays the existence of a forested wetland drained by placing wood boxes, clay tiles, and plastic pipe in the ground on the Merlin Spencer (deceased) farm in Menifee County, Kentucky.

Dry Level Ground

Many wetlands were eliminated by filling with soils left over from digging ditches and leveling high spots in fields. Most natural wetlands contain deep pools and small shallow pockets of water created by fallen trees and scouring during flood events. One would expect to see evidence of these features on sites that used to be wetlands. Level ground consisting of soil high in clay should be wetland. Locating a level or gently sloped area of ground containing soils that are high in clay without wetland characteristics often indicates a drained wetland. The lack of depressions and rough unplanted areas in a field provide strong evidence that land leveling and filling took place. A level field or woodlot should be examined with the same critical eye as that of a publisher examining a retouched photograph of a fashion model, realizing that perfection does not exist in this world.

Straight Streams

Streams that are straight or have few curves and bends were probably formed that way by people working to create large, unbroken fields. Naturally meandering streams and wetlands go together, as do straightened

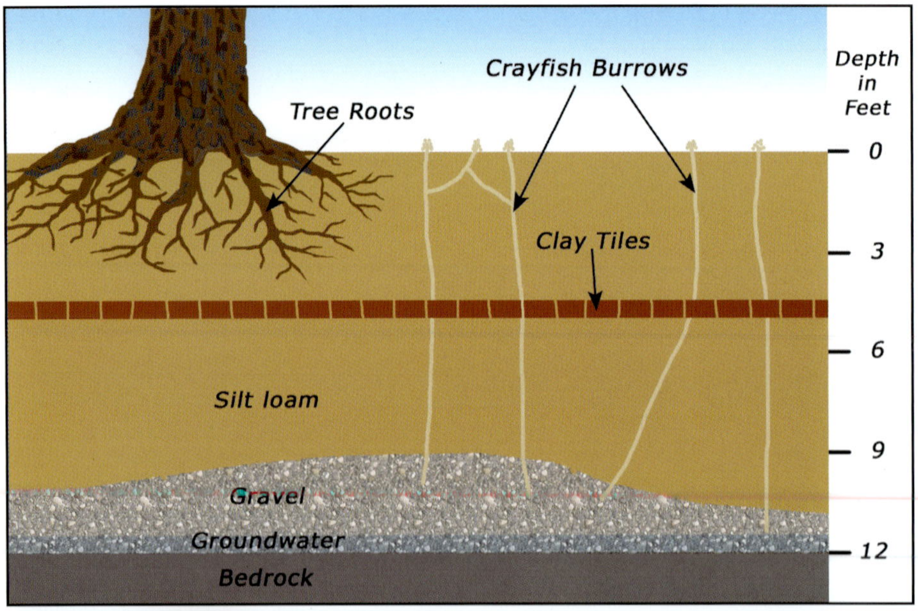

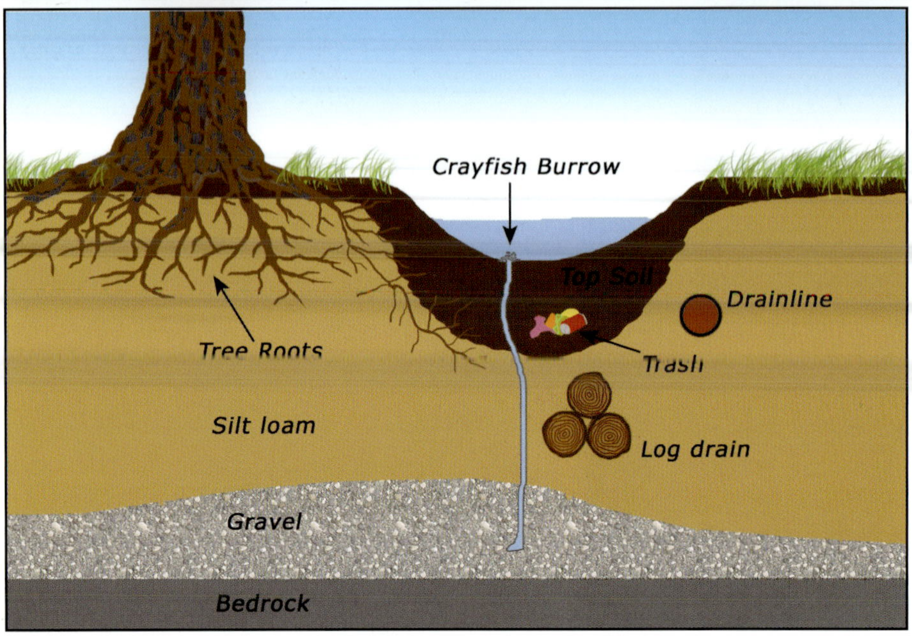

Water moves underground in a drained wetland (top left)

Center left image shows an attractive stream that is actually a constructed drainage ditch

Drainage structures are often buried under the drainage ditch (bottom left)

streams and dry lands. Expect natural wetlands to have been eliminated from an area if the streams were moved or channelized.

The consequence of a stream being straightened is generally the lowering of the water table on adjacent lands, which often drained wetlands maintained by an elevated water table. A straightened stream functions the same as a ditch dug deep enough to lower groundwater and dry soils for planting. John Johnstone, in 1808, described how groundwater-supported wetlands, originally formed by a river changing its course, could be drained by deepening and widening the adjacent bed of the river.[19]

A standard method for draining wetlands in mountains involved moving small streams that flowed off hillsides into open ditches constructed to traverse the shortest distance from the base of the hill to the main creek. Runoff from each hollow was directed into its own straight ditch that was made to enter the main creek at a right angle. Sometimes, there are dry basins still visible that show where these wetlands were located; however, in most cases, they have been filled and leveled.

Moved Creeks

Where farmland was created by moving creeks and draining wetlands, evidence of man-made changes can remain for hundreds of years. Recognizing historic modifications to creeks can help identify opportunities for ecosystem restoration. The following factors indicate where creeks and rivers were moved and channeled, and also signify where wetlands were destroyed:

- The creek is straight with few meanders.
- The creek follows the base of a hill

or a mountain for a considerable distance before crossing over to the other side of the valley where it continues to flow along the base of a hill.

- A railroad has been built in the same valley.
- A road has been built in the same valley.
- Agricultural land is present along the creek.
- Old fields or evidence of historic agricultural land is present along the creek.
- Continuous areas of level land are found along one side of the creek, indicating that agricultural fields were made larger by moving the creek.
- An abundance of high, vertical, eroding banks is found along the creek.
- Large trees, including their root masses, are falling into the creek.
- The slope of the creek channel is the same or steeper than the corresponding slope of the valley.
- There are few pools in the creek, and those pools present are shallow.
- The stream flows over sections of bedrock.
- Vertical drop-offs, known as head-cuts, are present in the stream channel.
- The creek channel carries flow after heavy rains, with waters spreading across the floodplain only after the severest of floods.
- Ditches enter the creek.
- Smaller streams enter larger streams at right angles.
- Pipe outlets are found along the creek.
- Logging or mining took place upstream, making it possible that the creek was used as a road for hauling logs and/or minerals.
- Small woody debris and leaves are uncommon in the creek.
- Exposed gravel layers are found above the current elevation of the creek, and examination of vertical banks along the creek shows gravel layers perched above the creek.

Slabcamp Creek in Rowan County, Kentucky was moved to the base of the hill by using a mule and plow in the 1800s. The moved creek was used as a road. The deep channel continues to act as a ditch, drying adjoining lands, though the fields have grown up to trees.

This channeled creek along the Housatonic River in Massachusetts has a straight path, steep high banks, few pools, and little woody debris, all characteristic of ditched streams.

This levee was piled along Brushy Creek in Greenup County, Kentucky after the stream was moved along the base of the hill to dry a wetland, creating a larger field.

Dry Depressions

Observing a depression of fine-textured soils that does not contain surface water or that lacks hydric soils can indicate a drained wetland. Dry basins that were once wetlands can range in size from less than one-quarter acre to as much as 300 acres. The surface and soils in these low areas is generally kept dry by buried drain lines, or by vertical drainage wells, which in turn indicate a drained wetland.

The topsoil depth in a depression can provide strong indication of a filled wetland. If topsoil depth on surrounding higher ground is only six inches deep and the topsoil is 30 inches deep in the depression, soils from surrounding ground were probably pushed downhill to fill the wetland.

A dozer removes four feet of topsoil used to fill a wetland at Bates Elementary School in Louisville, Kentucky.

Drained peat lands can also appear as dry depressions on the landscape. A number of authors discuss how much settling can occur when a bog is drained. Bogs were often dried with underground drains made of clay tile, wood, or stone.[20]

Finding an area of dry, fine-textured soil where a natural levee parallels a stream often indicates a drained wetland. Natural levees often act like dams, forming wetlands between the stream and the hillside. Land owners buried drain lines in these areas so that they passed under the natural levee to transport water directly to the stream, thereby draining wetlands.

Sinkholes

One would think that finding a wetland on top of a cave would be as rare as stumbling across water in the middle of a desert. But the heavy clay soils associated with karst, or limestone, topography can be expected to provide conditions for supporting a variety of wetlands over large voids in the ground.

Gritter Ridge is in the heart of eastern Kentucky's cave country. The mountains are steep and ridges narrow, with vast expanses of hardwood forests dominating the landscape. Here streams disappear from sight, sinking into the ground where they flow through miles of interconnected passages. In the early 1900s, Rufford Knox toiled to grow crops on a small ridge-top farm in Powell County. His land contained four large sinkholes, depressions in the ground that lack a stream outlet. Sinkholes can form when rooms in a cave collapse, and these had formed above Becky Cave, which underlies his farm.

One of the sinkholes maintained a year-round pool of water about two feet deep and one-half acre in size. Gary Knox, Rufford's grandson says, "It was like a large bowl; there was no way water could get out." In 1917, Rufford and his brothers improved the situation by using dynamite to blast a crack in the bottom of the sinkhole. Their action drained the wetland so that they could grow corn and tobacco in the depression. Later, Gary was able to use a lawn tractor to keep the area mowed.

Gary says that after extended periods of rain the sinkhole can hold water for short periods of time, and sometimes long enough for frogs to lay their eggs. That is when it is common to glimpse bitterns, snipe, herons, and vultures using the site; he suspects these predators are feeding on the stranded tadpoles. In 2008, while mowing, one of his tractor tires became stuck in a small hole in the bottom of the sinkhole. Today there are no visible holes or cracks in the bottom of the depression that indicate its drainage history.

Occasional emergent and ephemeral wetlands are found today in limestone sinkholes across Kentucky. However, most sinkholes are dry. While exploring caves on the Daniel Boone National Forest, it is common to enter rooms with a collapsed ceiling. These rooms often contain sharp, broken chunks of rock with waters pouring down from the surface, possibly indicating where explosives were used to drain other wetlands into the cavities below.

Soils

Existing wetlands often have gray- or black-colored soils, while wetlands that were effectively drained years ago have brown-colored soil. Wetlands that are partially drained may contain places with gray-colored soils. Buried drain lines and deep drainage ditches are effective at removing surface and groundwater from wetlands. This action, combined with plowing, serves to aerate soils, where ferrous iron that is present in the soil will oxidize and give soils their brown color. While the presence of gray-colored soil and a high water table can indicate a wetland, their absence does not prove the site was never a wetland.

Since ferrous iron is highly soluble, it can readily diffuse and return to aerated soil from moving water, as in runoff and groundwater.[21] A number of examples are given by soil experts on how ferrous iron can quickly oxidize, sometimes resulting in a visible color change within 30 minutes of exposure to air.[22] Try collecting samples of saturated gleyed soils from the bottom of a wetland and placing them in plastic containers holding some air. Within a couple of days, the exposed surfaces on these soils will probably turn brown, showing that once aerated, gray-colored soils can change color.

In the publication *Wetlands Losses in the United States 1780s to 1980s,* Dahl states that data on hydric soils has been used in some instances to approximate wetland acreage in the United States. "Because soil characteristics change slowly, even following drainage, summation of the soil acreages indicative of wetland conditions should approximate the wetland acreage at the time of settlement."[23] No further definition is given to the word "slowly", which has led many to believe that drained wetlands will retain a gleyed appearance with mottles for hundreds of years. This belief that soils will only change color after drainage at a slow, almost immeasurable rate has been adopted by some professionals to the point that they will only consider locations eligible for wetland restoration if they still contain hydric soils. Unfortunately, using hydric soils as a criterion for the selection of restoration sites would exclude the majority of successfully converted wetlands, as it is only in those wetlands that were poorly or partially drained where hydric soils remain. Individuals who drain wetlands professionally report that, once successfully drained, the gray-colored soils found in wetlands will turn brown.[24]

Looking Past the Trees

Features of drained wetlands such as ditches, straightened streams, and dry depressions are more difficult to see once an area has grown up to trees. Looking for drained and filled wetlands in wooded areas will greatly increase the number of projects available to the land manager. From 1982 to 2002, the quantities of forest land in the United States increased by 1.9 million acres, while the acreage in crop land decreased by 52 million acres.[25] Since the majority of wetlands were originally drained to create crop land, and currently large numbers of these fields are growing up to trees, it is important to consider forested areas when looking for wetland restoration sites.

Signs of drainage remain in this old field growing up to trees for 80 years. The rock pile from land clearing and the straightened creek at the base of the hill show how the land was changed for farming.

Tree Plantations

Tree plantations may indicate the presence of relic wetlands. Shallow ditches were often used to remove surface waters in order to provide a seed bed for growing more commercially valuable pines throughout the south. Most of the ditches used to drain areas for pine are not maintained and are now hidden by trees and shrubs, yet continue to function. The shallow ditches used to drain wetlands for pine can be spaced as much as one-half mile apart, and often empty into constructed ditches at least four feet deep. Roads were often built along the edges of these deeper ditches using the soils recovered from creating the ditch. Culverts were placed beneath the constructed roads to transfer waters from the shallow ditches on the other side into the deeper channel.

Trees were also planted in wetlands by directionally plowing saturated soils to create "lands". This practice, often called "bedding" in forestry, provided a series of long, narrow ridges whose top rises above surrounding water. Pines planted on these ridges will survive, despite standing water, and as they grow, will dry the surrounding soils.

What was common on many national forests was to clear low stocked forested stands by dozer to prepare areas for planting. Shallow ephemeral wetlands were often drained during this process to create additional areas for planting, while other wetlands were simply filled, because they were considered unproductive and provided convenient openings for disposing of brush and debris.

Puddles of Water

The deepest portions of drained wetlands can be revealed by a heavy rain. Evidence of historic wetlands is often shown by puddles of standing water that disappear in a couple of days. Buried drainage systems overwhelmed by flooding yield a brief glimpse of how large and how deep a wetland was prior to modification.

Road Ruts

Ruts in a road can be a sign that the road was constructed over a historic wetland. They generally indicate where soils high in clay are compacted to the point where they hold surface water. Often the only sign of a small wetland impacted by vehicle traffic is a deep puddle of water in the road.

Road ruts generally contain water as long as the road is used, as the regular traffic keeps soils in the rut compacted. Once closed to vehicles, road rut ponds can be expected to become drier as roots, combined with freezing and thawing, will break up the thin layer of compacted clay in the bottom.

Road ruts containing water are also formed by vehicle traffic over springs and across ephemeral streams. These puddles can have a rocky bottom. A spring-fed road rut can be identified by the presence of saturated soils uphill from the puddle, water seeping from the face of the cut bank above the road, and water flowing from the puddle down the road when it is not raining.

Ruts in fields often indicate where drain lines buried to eliminate wetlands have been damaged by tractors or trucks when the soils were moist. Damage to drainage systems caused by rubber tires may allow portions of the original wetland to return. These wet places are generally quickly repaired on an active farm. An abundance of ruts can also show where an existing wetland has been damaged by OHVs.

Two strands of barbed wire barely visible on the tree indicate that the forested area was once farmland. Evidence of agriculture usually corresponds with a landscape that was drained for crop production.

Neighboring Lands

While one farm can have absolutely no wetlands, neighboring farms may have any number of them. This situation can be due to variations in work ethics or how often a landowner participated in government-sponsored drainage programs. The presence of wetlands often changes based on ownership. If one observes the land surrounding the site upon which one is working, and that land has wetlands, while the location one is working does not, the site one is working has probably been drained.

Verbal History

Talk to residents who really know the land, especially those who raised crops in an area for years. Ask if they know of places that once had puddles of water, or that stayed wet into late summer. Inquire about the location of "swampy" places, as the current definition of "wetlands" is unclear to many. A place where they have often gotten their tractor stuck indicates a drained wetland.

One does not always think of wetlands being drained in mountainous areas, but older residents who lived in the hills remember how common these wet places were. Merlin Spencer remembered his grandfather laboring to farm steep hillsides above Beaver Creek in Menifee County, Kentucky. "Those mountains were tough to plow, lots of stumps to go around, and there wasn't any topsoil," he said. "My grandfather had to drain the swampland along the creek or we all would have starved."[26]

Written Records

It is rare to find written records of drainage practices on a farm. Those who did this work were often too busy to record these procedures. Do not expect to find many reports of drainage activities in government offices either. Most wetlands were drained so many years ago, that there is no written or verbal record of their presence.

The best opportunities to find documentation of drainage activity are those funded by the government. The United States Department of Agriculture's Natural Resources Conservation Service (NRCS), formerly the Soil Conservation Service, has actively designed and funded drainage activities for most of its existence. Detailed plans were often prepared for drainage projects, and can be found on file in NRCS offices. However, these files are regularly updated and purged after a number of years, and may be difficult to locate once moved for archival to central offices.

NRCS offices maintain records of farmland in their service areas where certain parcels are labeled as "PC" for Prior Conversion. This designation was given to drained wetlands being farmed prior to passage of the 1985 Farm Bill. While the designation of PC can help identify the location of drained wetlands, it should not be considered as the complete documentation of where wetlands were located. The NRCS would only make the PC determination for those individual fields proposed for enrollment in USDA programs at the time, which means that other drained wetlands, even those being cropped on

The faint lines on the LiDAR image show narrow and shallow drainage ditches and small moved creeks in fields last farmed in the 1800s. The image was derived from Hillshade 1ft x 1ft resolution LiDAR data. The area is located at the confluence of Indian Creek and Hunting Creek, in Newberry County on the Sumter National Forest in South Carolina.

Ditches and moved creeks are difficult to detect on this 2005 USGS color infrared Digital Ortho Quarter Quad with 1m x 1m resolution. The pink color shows large pine trees, the gray and blue, large hardwood trees.

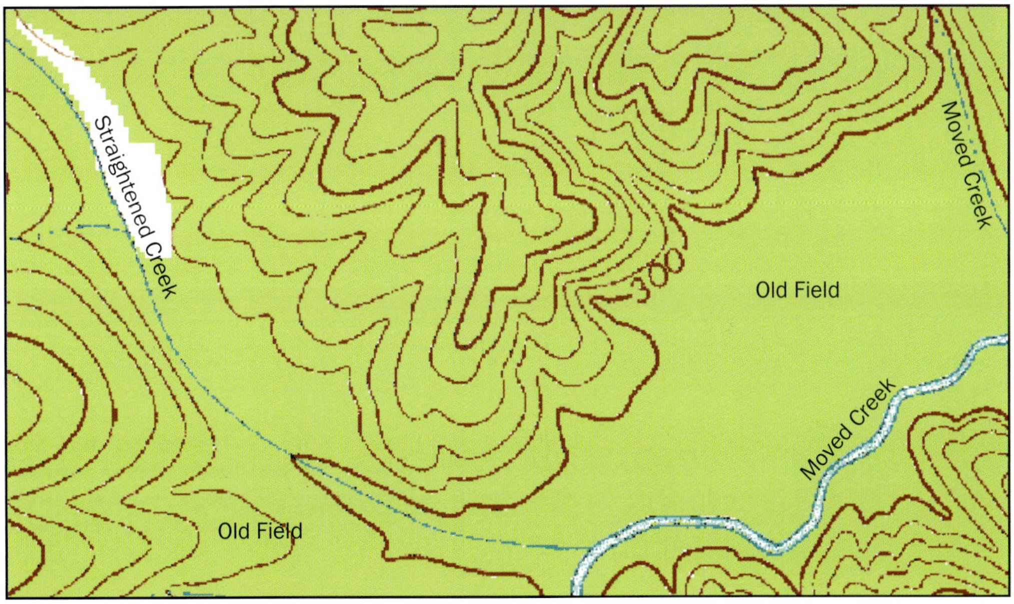

Larger perennial creeks that were straightened and moved can be seen on this 1:24,000 scale map series, 7.5' USGS topographic map with 10ft contour intervals.

Scale for each image is 1:11,000 Prepared by Jason Jennings and Andy Maceyka.

the same farm but not enrolled in the USDA program, may not have been recognized. Because many farmers did not participate in USDA programs, it is not uncommon to find tracts of land containing drained wetlands that do not carry the PC label.

Aerial Photographs

Aerial photographs can help show where drained wetlands occurred on a landscape. The straight lines of ditches are often visible on aerial photographs, especially when an area has not grown up to trees. Observing a series of parallel straight lines in a small area can indicate where ditches were used to drain a wetland. The practice of "lands" used to farm wetlands by plowing to create a series of parallel ridges can be seen on aerial photographs. An infrared aerial photograph may show the presence of buried drain lines in northern areas, if taken when frost is coming out of the ground. However, after the land is plowed or is growing vegetation, these buried drain lines are generally not visible.[27]

Landscape patterns can provide clues to the historic presence of wetlands in an area. Aerial photographs clearly show the outline of square-mile units of land, known as sections, bordered by roads in the agricultural

This 1978 color infrared aerial photograph shows a landscape near the Licking River in Bath County, Kentucky where a large number of wetlands have been modified for farming.

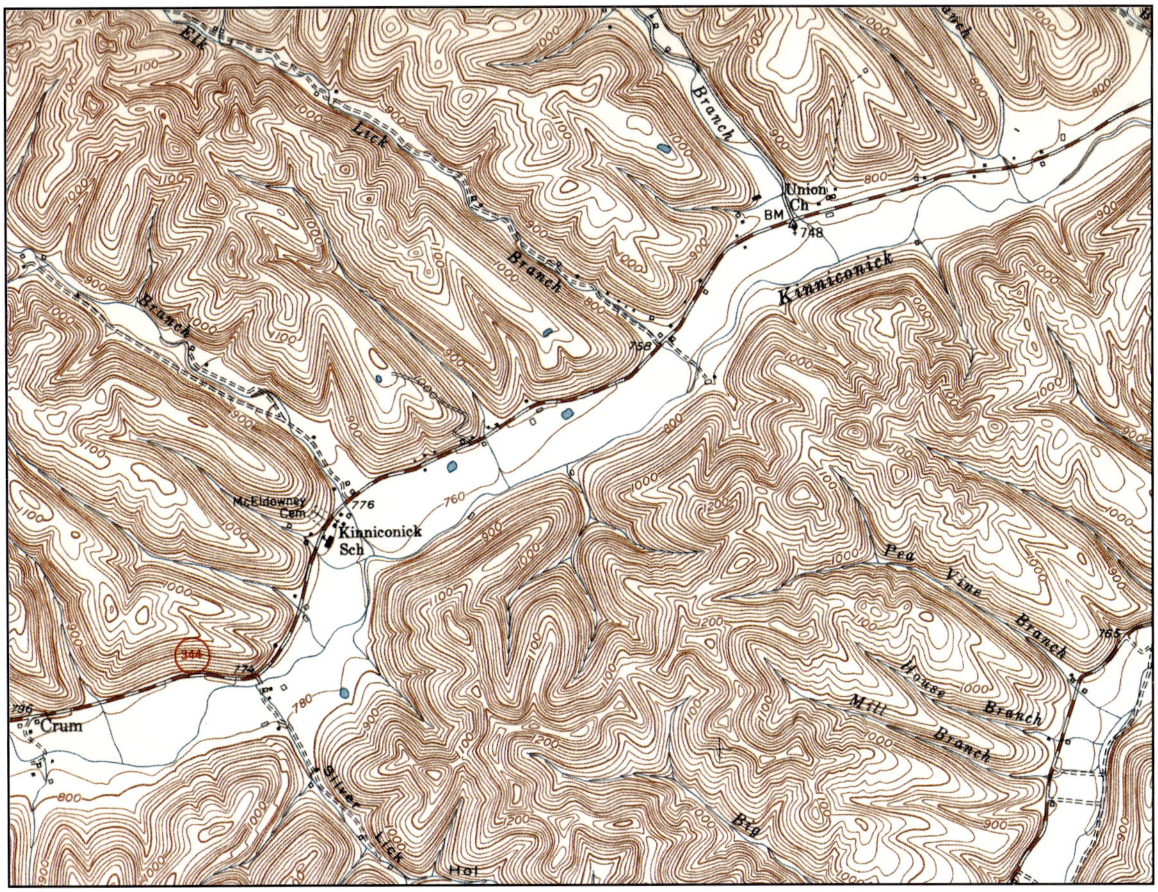

Kinniconick Creek was moved so it would flow along the base of the hill to create farmland. All of the tributaries were straightened and changed so they would enter Kinniconick Creek at right angles to create large and dry fields. The contour lines are at 20-foot intervals on this USGS topographic map showing a portion of Lewis County, Kentucky.

region of the midwest. It is not difficult to find sections that contain wetlands adjacent to those that do not. These sections that contain wetlands can easily be seen from an aircraft at an elevation of 35,000 feet when flying over the prairie pothole region in South Dakota and Minnesota. Sections that lack wetlands often indicate farms where landowners have taken action to remove them. By looking closely at those sections that do not have wetlands, the darker colored imprint of eliminated wetlands can often be detected in the agricultural fields.

One can examine aerial photographs taken of the same area over different years to identify drained wetlands. Historic wetlands are often more visible when comparing one year to another. Matching recent aerial photographs with photographs taken years ago can help distinguish drained wetlands. The presence of shallow water and uncultivated areas surrounded by fields may indicate wetlands. In some areas, it is possible to find aerial photographs taken in the 1930s. These old photographs may be stored at USDA Natural Resources Conservation Service, USDA Forest Service, and other government offices.

Florescent Flags

Farmers often mark the location of surface inlets in their fields so that they can avoid damaging them while planting or harvesting. Generally using florescent orange flags atop long poles, these markers are readily seen in the bottom of large and small depressions that were previously wetlands. Water enters the surface inlet near the marker, and is then carried beneath the ground via a buried drainage system to an open ditch or creek some distance away. The flags used to mark surface inlets can be spotted on both high and low ground.

This hand-dug ditch, which meets at right angles with the straightened stream, shows where a wetland was drained.

The Basics of Wetland Restoration

Chapter 3

Professionals often downplay the value of creating wetlands, and actually discourage the practice of building them. Their reasons center largely on the failure of these projects to hold water as planned, and on how artificial they can appear. Both wetland creation and restoration projects can fail because of improper construction techniques, but fortunately, both practices may consistently produce wetlands that look and function like natural habitats.

It is possible to restore and to create wetlands that will be similar in appearance and function to natural ecosystems. Those who are new to wetland establishment often become concerned about determining whether the actions they are proposing fit into the realm of wetland restoration or should be considered a wetland creation. This can cause much frustration as, in many cases, one can no longer tell if a site was a wetland. Many historic wetlands have been so changed by previous drainage activities and vehicle damage that few, if any, signs remain to show the area was once wetland, let alone what type of wetland it might have been. Fortunately, when it comes to the implementation of wetland restoration practices on the ground, how one proceeds with a wetland restoration or creation project is much the same.

Locating Drained Wetlands & Sites for New Wetland Construction

What is critical to recognize is that someone before has managed the area now being considered for wetland restoration, and in most situations, his or her livelihood depended on that land. He or she had to figure out how to remove excess water to feed a family, build roads and railroads where creeks and swamps were in the way, and prepare ground for home construction. What may be sobering to realize is that he was every bit as smart as those working on wetlands are today. Therefore, those restoring these sites need to work just as hard to disable the actions they have made, sometimes over generations, to dry the land.

Electric transmission line corridors offer excellent locations for building emergent wetlands that will not grow trees, which can help reduce maintenance costs.

This book will describe techniques for building a wetland on almost any type of landscape. There are many locations where wetlands can be created: mountain ridge-tops, benches, small bottomland fields, mined areas, old roads, and even transmission line right-of-ways (ROWs).

Fields

Many fields can be made into wetlands, even if they must occur on a slope by stepping down multiple shallow pools. Creating a wetland in an open space is one of the lowest cost wetland projects to complete because of the easy access and the lack of trees and shrubs to be removed.

This emergent wetland was restored in an old field on the Daniel Boone National Forest in Kentucky.

Mined Areas

When searching for a place to build a wetland, locations where soil and rock have been removed should not be overlooked. Abandoned gravel pits might be unattractive, but can be converted into beautiful wetlands. Advantages to building wetlands on previously mined sites can include cost savings by being able to avoid or use abbreviated *Archeological Surveys*, *Endangered Species Surveys*, and *National Environmental Policy Act* (NEPA) documentation. Society generally supports

A small ephemeral wetland (above) was built in 2008 in a tree gap by using a large dozer (Case 1150E) to shape and compact soils high in clay.

This emergent wetland (left) was constructed on an area where clay soils were removed to build a flood control reservoir in eastern Kentucky.

This small emergent wetland was created on a coal mine in Whitley County, Kentucky.

This small ephemeral wetland was constructed two years ago in a closed logging road and is used by wood frogs for breeding.

This is a two-year-old wetland in a closed logging road on the Monongahela National Forest in West Virginia. It provides drinking water for bats and breeding habitat for wood frogs and spotted salamanders.

building wetlands on mined areas, as most will appreciate efforts to improve an unsightly eyesore.

Sites where gravel has been mined can often be shaped into wetlands. Generally, large quantities of soil and rock have been removed down to the elevation of the water table. Sometimes removing additional gravel can help provide the water depth needed for a wetland. In many cases, a barter system can be arranged, wherein a contractor in need of the gravel can be persuaded to create wetlands in exchange for the rock.

Sites where mineral soil has been removed for road construction can make excellent wetlands. It only takes a limited amount of fine-textured soil to create a wetland on these areas that have been scraped down to clay and bedrock. Look for topsoil that has been piled around the edge of a borrow pit that can be spread over the wetland when it is completed.

Small Forest Openings

Wetlands can be built in small openings in the forest, sometimes called tree gaps. A small dozer or excavator can be used to build a naturally-appearing wetland in a space as little as 40 feet in diameter. In mountainous areas, small wetlands can be built on narrow benches and on ridge-tops in tree gaps. A small dozer the size of a John Deere 450 can be walked for miles off a road into remote locations to build small wetlands by going around and between, not over, larger trees.

Timber Harvest Areas

Logged areas offer great opportunities for wetland establishment. When working on national forest land, the costs associated with the planning of wetlands can be lowered by designing sites in areas where trees have been cut and sold in recent years. In many cases, required *Archeological Surveys*, *Endangered Species Surveys*, and *National Environmental Policy Act* documentation have been completed for timber sale areas. Roads that were built to haul trees often provide easy access to potential wetland construction sites. Skid roads can also be prime locations to construct a number of smaller, linear wetlands.

When walking around areas where trees have been harvested, one should watch for places where skidders and trucks have sunk into soft ground and left behind ruts. If these ruts contain water, this is a strong indication that soils are fine in texture and can be shaped to form wetlands.

One should use care when marking wetland construction sites in timber sale areas where large stumps are found. Large diameter stumps from recently cut trees are difficult and costly to remove. Consider building clusters of wetlands in areas where trees were cut from 10 to 15 years ago. Older stumps no longer pose a problem for removal, and the dense clumps of saplings growing around the old stumps can be easily cleared by a dozer.

This series of small wetlands was constructed in a skid road on a slope to control erosion and to close the road to vehicle traffic.

Old Roads

Wetlands can be built in roads that are no longer needed. In many areas, small, linear-shaped wetlands can be built at a reasonable cost in roads that are not surfaced with gravel or asphalt. Wetlands are now being built in roads to create barriers that trucks and OHVs cannot cross. Ruts found in roads that contain water or are deep and muddy indicate where soils can be shaped and compacted to form wetlands.

Transmission Line Right-of-Ways

Transmission lines right-of-ways are common features that cross fields, forests, streams, and wetlands on both private and public lands. Utility companies are always looking for ways to reduce woody plant encroachment in their ROWs, and to show customers how they are helping the environment. Wetlands in ROWs can reduce maintenance costs, as woody plants will not grow in water or saturated wetland soils. Wetlands can help sequester carbon from the atmosphere. Building them in ROWs may be an effective way of using the non forested areas to reduce global warming. East Kentucky Power Cooperative, Inc. and American Electric Power Company have financed the construction of over 150 wetlands within the ROWs on the Daniel Boone National Forest in Kentucky since 1990.

This attractive emergent wetland was constructed in a transmission line ROW at Tualatin Hills Nature Park in Portland, Oregon.

Golf Courses

These intensively managed landscapes offer tremendous opportunities for wetland restoration. Adding wetlands of different depths and sizes can enhance the beauty of a golf course, create opportunities for wildlife viewing, and increase golfer interest. Unlike gardens, which require frequent watering and weeding, wetlands are basically maintenance-free after they are built. Goose problems associated with water and wetlands can be discouraged by not building islands that can be used for nesting, and by not mowing near the water's edge.

Soil Surveys

Soil surveys have been prepared by the NRCS for almost every county of the United States, and provide a wealth of information for planning wetlands. The soils found in each county are described according to soil series that include characteristics such as texture, depth, and elevation, and are available on the web: http://websoilsurvey.nrcs.usda.gov. Soil surveys also contain aerial photographs overlaid with soils mapped according to soil series and slope. The mapping units allow one to quickly scan areas for the presence of soils that may be suitable for wetland construction.

Soil surveys contain tables that can assist in evaluating the potential of an area for wetland construction. The tables often include information on the elevation of the water table, presence of subsurface sand and gravel layers, and even where cracks may occur in the bedrock. Sometimes it helps to use the information in the tables in the reverse manner of which it was intended. For example, a soil series where agricultural drainage is recommended, and soils are not suitable for septic drain fields because of slow permeability, and a seasonally high water table, also describes locations one may wish to examine for wetland restoration. Along the same lines, soil series where few problems are expected for farm pond construction can also be expected to be suitable for wetland construction.

The information contained in the soil survey books should serve as a guide, not an authority for selecting locations for wetland construction. It is entirely possible to find sites that are suitable for wetland construction within soil mapping units that indicate otherwise. On the other hand, areas not suited for wetland construction may occur within mapping units that state that they are. The presence of these small areas, known as inclusions, were generally not identified during the original soil survey, demonstrating the reason it is important to personally test soils in an area when planning a wetland project.

Indications That an Area May Support Wetlands

- Wetlands already established in an area are a sure sign that additional wetlands can be created or restored. The current wetlands show that soil textures and the water table are suitable for building wetlands in nearby areas.

- Constructed farm ponds are a good indication that soil textures are suitable for constructing wetlands. A farm pond shows that soils at that elevation may have enough clay to be shaped into wetlands.

- Water standing in roadside ditches that also contain cattails is evidence that wetlands were drained in the area, and that soils are suitable for their reconstruction. These mini wetlands often are indicative that much larger wetlands were once present on the landscape. Willows can be a sign of a seasonally high water table, often present on drained wetlands.

- White oak trees growing on mountain ridge-tops often show that clay soils will support wetlands. In Virginia, willow oaks can show where clay is located. Big leaf maple trees appear to indicate clay on Vancouver Island, British Columbia. Sweet gum trees are also good hints of the silt-clay-loam soils on which many wetlands were once found.

- Ruts in roads indicate where soils can be shaped and compacted to hold water. The presence of road ruts shows where wetlands can be dug in and near dirt roads.

Building in the Sun or Shade

Whether the wetland is built in the sun or in the shade can have a big influence on the plants and animals that will use the wetland. Wetlands constructed in sunny versus shady locations can be expected to show the following characteristics:

Sun

- High diversity of plants
- Nesting habitat for geese, wood ducks, mallards, pied-billed grebe, red-wing blackbirds
- Feeding areas for shorebirds
- Improved habitat for turtles
- Frog and salamander tadpoles develop earlier in warmer water
- Greater losses to evaporation mean that waters will dry sooner
- Possible problems with dense cattail growth and invasive species such as non-native purple loosestrife

Shade

- Feeding habitat for wood ducks, hooded mergansers, and black ducks
- Breeding habitat for wood frogs and woodland salamanders
- Watering sources for forest-dwelling bats
- Reduced losses to evaporation means that waters will remain longer
- Low diversity of plants
- Possible problems with non-native invasive plants, such as Japanese grass

Building a wetland in a forested area does not necessarily mean that the result will be a wetland that has shaded characteristics. Removing a large number of trees on a site can open it up to daylight, creating a wetland with sunny characteristics until trees grow and shade the water. These tips can help maintain shade around a new wetland:

- Build the wetland in a small opening between tall trees.
- Remove only shorter trees in the center when creating a clearing for construction.
- Use care to leave as many trees standing along the southwest edge as possible, so that the wetland pool will remain shaded in the afternoon hours.
- Construct the wetland so that it will be about the same distance across as the height of the surrounding trees.
- Build on a north-facing slope.

Building a wetland in a wooded area so that portions are in the sun and portions are in the shade provides conditions where characteristics of both types will be found, generally resulting in an increased variety of plant and animal use of the new wetland.

Site Selection

When looking for a place to build a wetland, search for land that appears level, whether in a field, young forest, or canopy gap within an older forest. An area that is 60 feet or larger in diameter will provide enough space for building a wetland that appears natural.

Some individuals assume that damming a perennial stream would be the easiest method of beginning to construct a wetland. However, no matter how small the stream may be, damage caused by heavy rains is a substantial risk, as well as possible other negative impacts to the natural surroundings, and potential large economic costs may be involved.

A mountain ridge is an excellent location for building a wetland. Much needed water for wildlife can be supplied, and one can avoid problems associated with flowing water and the buried gravel often found along streams. A wetland can be built at low cost by choosing a level area on high ground, away from streams, with few trees and

This team is using a construction level to measure slope. A laser level is operating on the yellow tripod. (Barry Nichols photo)

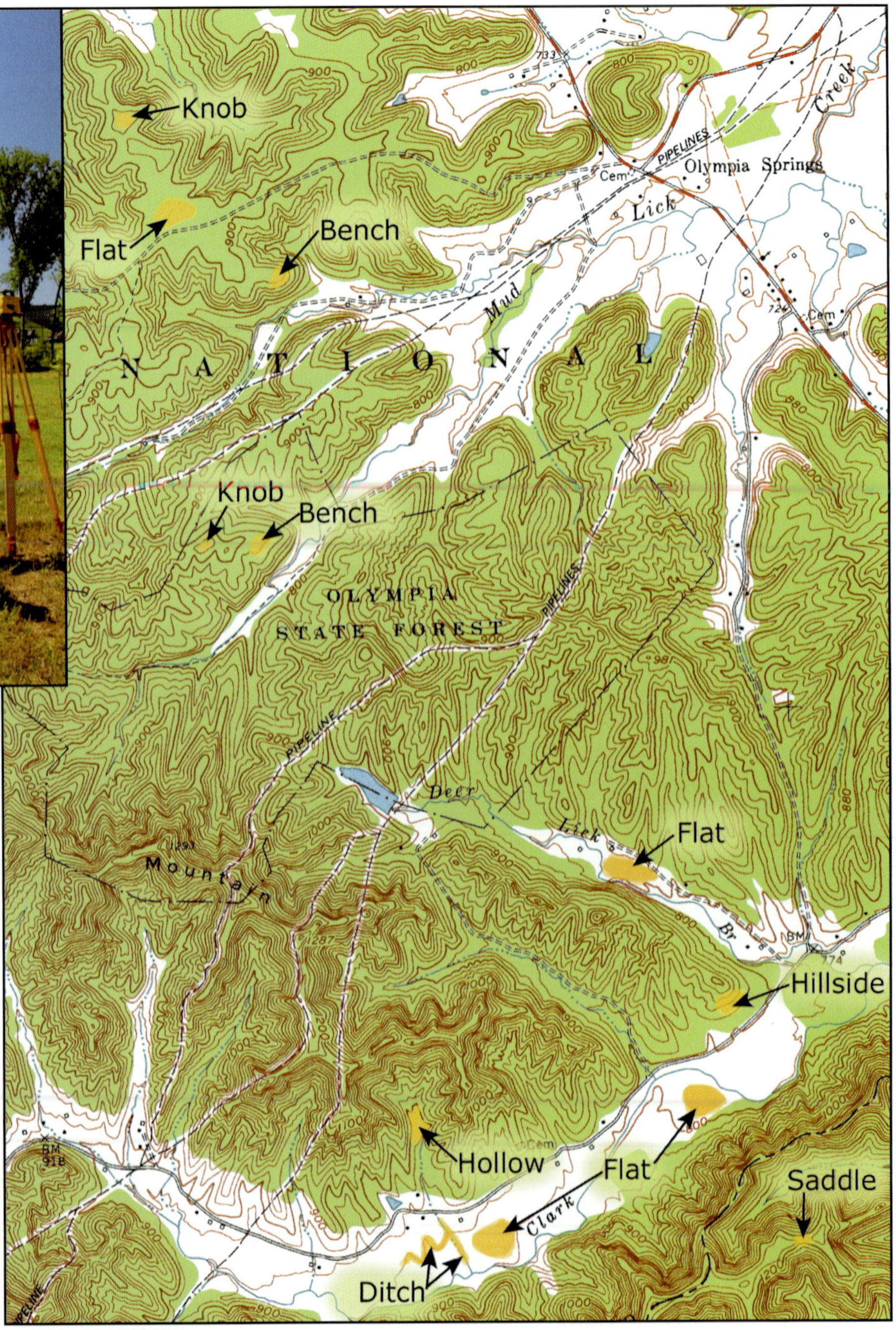

Possible locations for building wetlands (contour lines mark 20-foot changes in elevation)

soils that are high in clay.

Select a location with a slope of six or less percent, which is basically the same as an area that changes three or less feet in elevation from upper to lower edge along a 50-foot segment. If the ground looks flat enough to pitch a tent for a good night's sleep, then it is level enough to support a wetland. In mountainous areas, use a construction level, laser level, or clinometer to measure slope, as ground often appears flatter than it really is. If the area is steeper than six percent, it would be better to build several shallower wetlands that are stair-stepped down the slope than to build a large wetland with a high dam.

Mark the perimeter of a construction site that is at least twice as large as the surface area of water planned in the new wetland using flagging. The added space will be needed for piling woody debris, vegetation, and topsoil, and for building a dam with gradual slopes. Do not be afraid to remove some trees to create a wetland; in many cases, the reason the trees are currently growing on a site is because it was historically drained.

Flagging can be used to mark the perimeter of a wetland construction site. Remember that it is important to mark an area twice the size of the planned wetland to provide room for piling vegetation and topsoil, and for creating gradual slopes.

Wetland Construction Illustrations

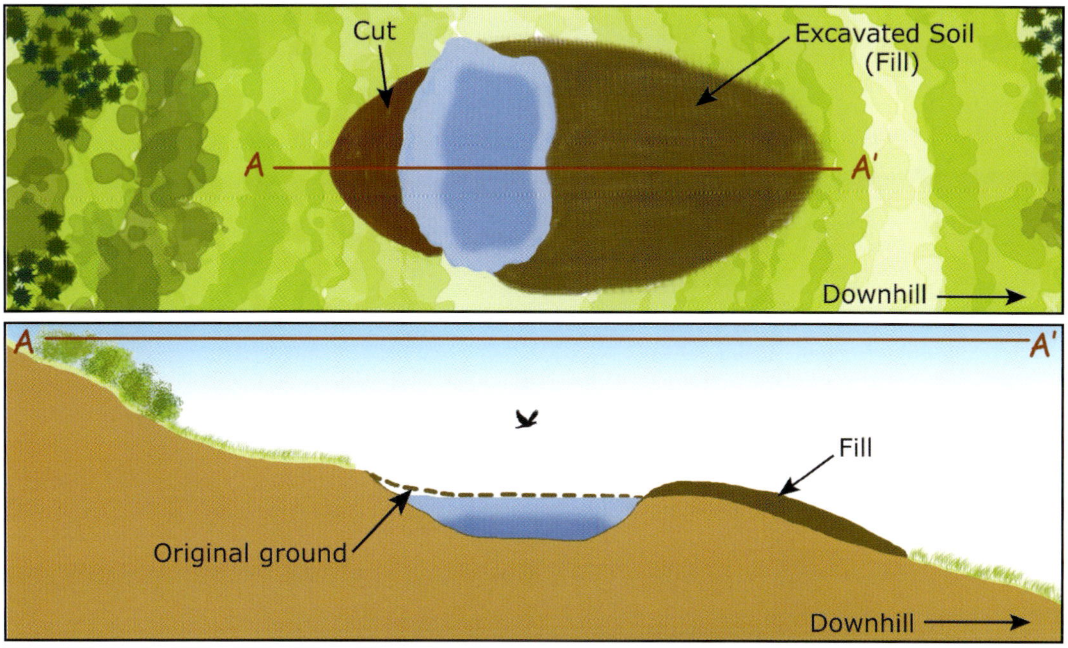

Bench location, Plan view (top); Bench location, Profile view (bottom)

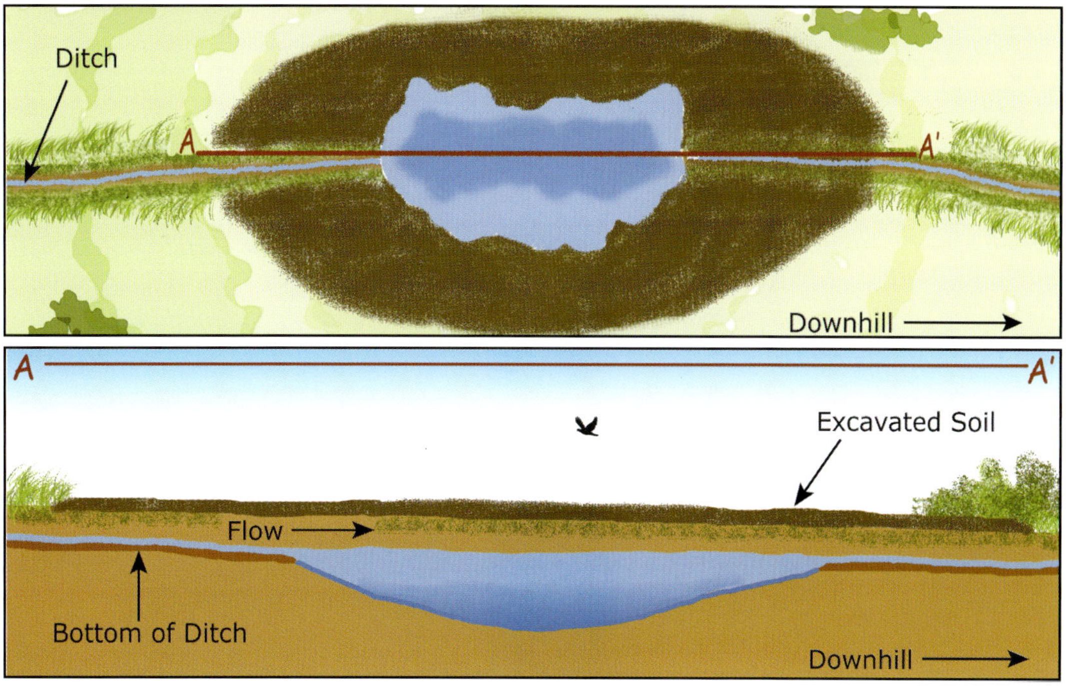

Ditch location, flow maintained, Plan view (top); Ditch location, flow maintained, Profile view (bottom)

This small wetland was constructed in a ditch where flow was maintained by not changing the elevation of the inlet or the outlet.

Before deciding to build a wetland in a valley, ditch, or small stream, it is best to walk both upstream and downstream of the site to identify any head-cuts present. A head-cut located upstream can fill the new wetland with soil. A head-cut downstream can drain the new wetland as it moves upstream and cuts through the dam and wetland. Unless action is planned to stop head-cuts above and below the site, a different location should be found.

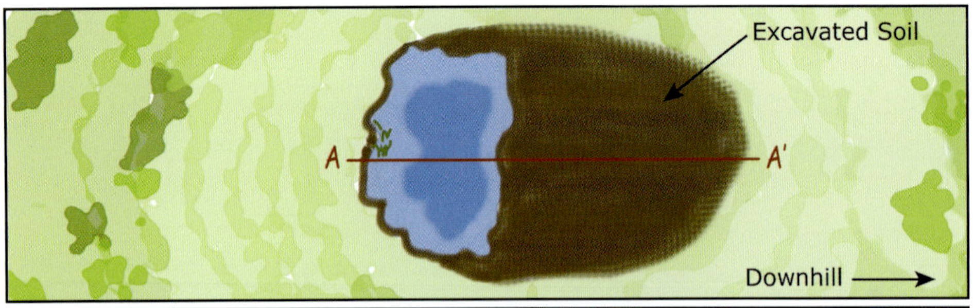

Ditch location, flow blocked, Plan view

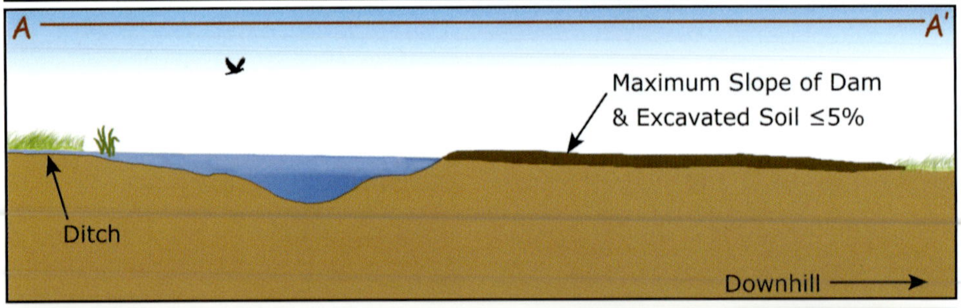

Ditch location, flow blocked, Profile view

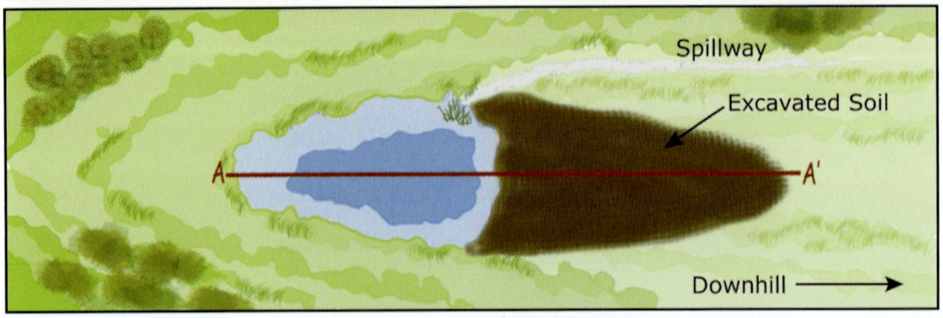

Hollow location, Plan view

Hollow location, Profile view

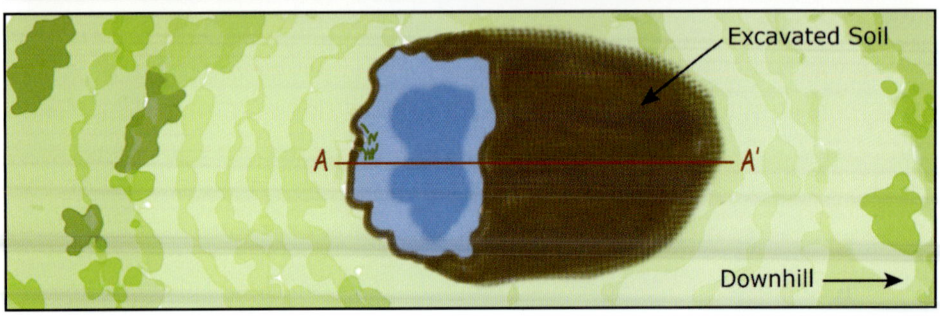

Flat location, Plan view

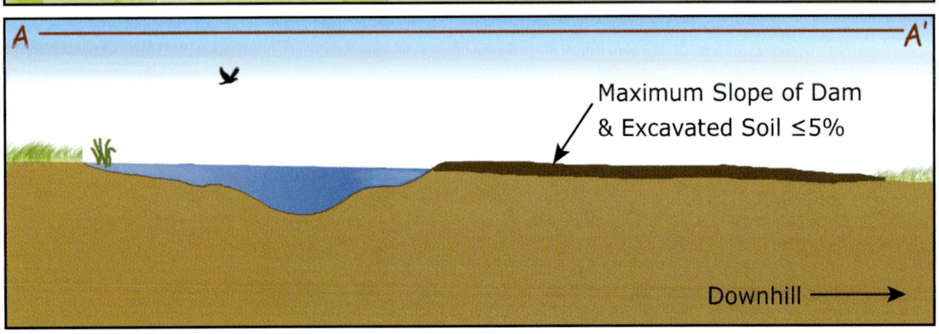

Flat location, Profile view

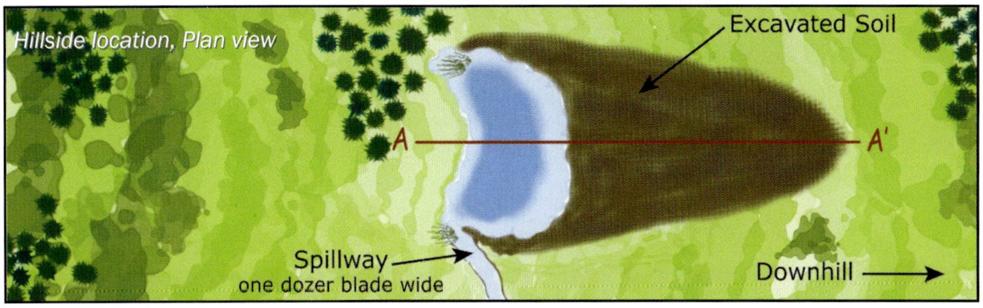

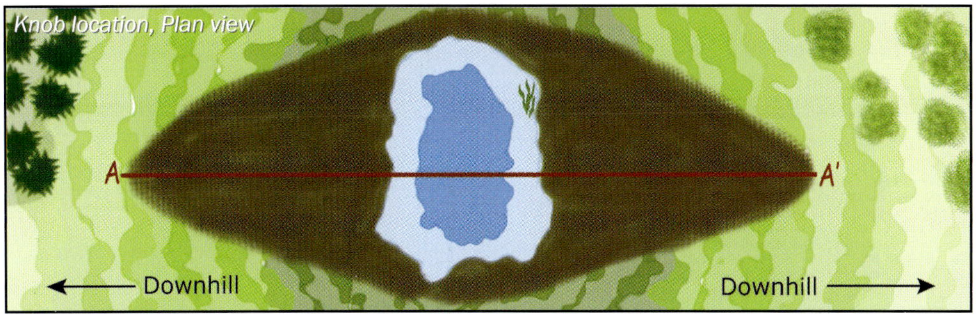

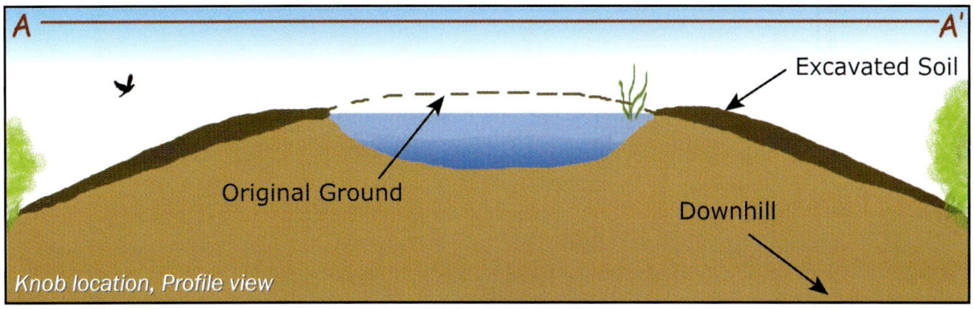

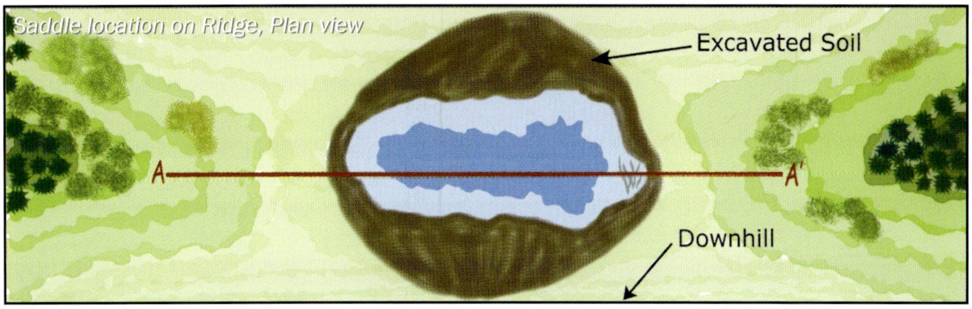

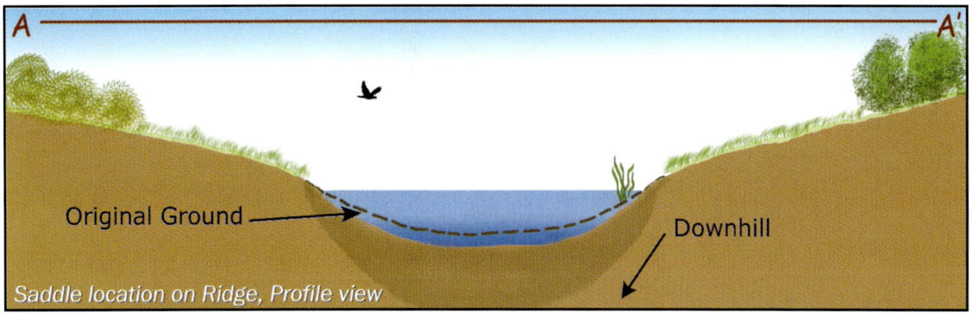

The degree of success will increase greatly when it is decided whether the wetland being built will be supplied primarily by precipitation and runoff, or by groundwater.

Other Considerations

Deciding Between Surface Water and Groundwater

Whether planning to restore or create a wetland, decide if it will be supplied primarily by surface water or by groundwater early in the process. A surface water wetland holds rainfall like a cereal bowl, within a depression made of packed soils that are high in clay, and a dam that serves as a rim to keep waters from flowing downhill. A groundwater wetland is like an old fashioned hand-dug well; it exposes a high water table that is often surrounded by soils high in sand or gravel. Different construction techniques are used to restore surface water and groundwater wetlands. A soil test provides the information needed to decide which strategy to use.

Use a posthole digger or soil auger to dig a hole at least three feet deep near the center of the proposed construction site. A 1.5-inch diameter soil auger attached to a four-foot long handle works very well to test potential restoration sites. A soil probe can be difficult to use in rocky or clay soils. Watch to see if water seeps into the hole from the bottom and sides. If the hole fills partially or completely with water, or the slurp of water is heard as the auger is removed, a high water table is present, and a wetland can be built that will fill with groundwater. Soil texture will not be a concern when building a groundwater wetland, as a depression is simply excavated to expose water contained in the soil. However, if water enters from the top of the test hole, if the hole is dry, or if only a little water seeps in, plan on building a wetland that is designed to hold surface water.

A tile probe offers a fast way to determine if groundwater is near the surface. The tool is a long, small diameter steel rod mounted on a handle that is easily pushed into the ground at possible wetland construction sites. While longer models are available, the four-foot version is the easiest to use. When removed from the soil, water drips off the tile probe, showing groundwater elevation, similar to the way a dipstick measures the oil level in an engine. The tile probe can also be used to determine if rocks are present in the ground, and if bedrock is near the surface. The tile probe is not a reliable tool for determining soil texture.

A seasonally high water table can be more difficult to detect in dry weather, but there are several signs that betray its presence. One is to look for gray-colored soil, and for mottling. Another is to allow more time for water to enter the test hole. A board can be placed over the top of the hole to prevent rain from entering, and one can return the next day to see if water is present in the hole. It is a good idea to test the site again when conditions are wetter to check if water is nearer the surface. Test holes showing a weak flow of groundwater, or where water is far below the surface suggest that a wetland excavated on the site will contain little water, and possibly for only short periods of the year.

If a high water table is not present, it is important to determine if the mineral soils located beneath the topsoil layer, which is dark-colored and containing roots and leaves, can be made to hold water. Soil textures that are suitable for holding surface water are high in clay and silt.

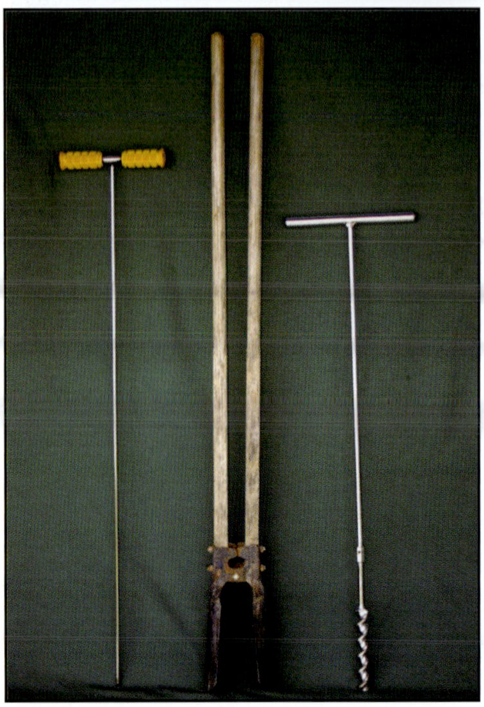

The tile probe (left), posthole digger (center), and 1.5-inch diameter soil auger (right) are useful hand tools for identifying the presence of groundwater, and when determining soil texture on wetland projects. (Sarah Biebighauser photo)

A simple test can be used to determine if the clay and silt content is high enough to build a surface water wetland. Make a one-inch diameter ball of soil, add some water, and then attempt to form a thin ribbon of soil between thumb and forefinger that is two inches long or longer. If a ribbon can be formed that is at least two inches long before it breaks, then the soil has enough clay and silt to be shaped and packed to hold surface water. Soils that contain adequate amounts of clay are firm and take some strength to reshape. Molding them is like playing with a Tootsie Roll; it takes time and gets messy. Soils with too much silt content feel like flour when rubbed; those with too much sand feel gritty; and those with just enough clay feel firm and smooth. Some think it easier to test for clay by rolling a walnut-sized chunk of soil between their palms after adding some water, and then forming it into a strand about the diameter of a drinking straw, and finding the length to which it can be worked before breaking. Sections two inches or longer provide strong indication that soils on the site can be shaped and packed to hold surface water.

Remember, the goal is not to see how long a ribbon can be made from the soil, but rather to measure the length of the self-supporting sections that break off. Mix only enough water into the soil so that it can be molded without dripping. Next, force the soil between the thumb and forefinger so that it rolls off the side of the finger in a thin ribbon. The lengths of the pieces that break off indicate clay content. Measure those sections that fall to the ground. If they are two inches or longer, the soils will have enough clay to reshape and compact to hold surface water.

Soils that ribbon in sections from one-and-a-half- to two-inches long can still be used to build surface water wetlands, but more exacting construction techniques should be followed along with a larger watershed for success. The soil moisture content on the worksite must be correct to obtain compaction. This can involve delaying construction until a wet time of year or continually spraying water on soils during construction. Equipment operators must also take extra steps to insure compaction by spreading soils in layers no greater than six inches, and then traveling over each layer at least four times for compaction. A machine such as a roller, tractor, or backhoe with rubber tires should be used to compact each layer of soil in the dam and groundwater dam. Additional

The long, thin, self-supporting ribbon shows that these soils have enough clay to be shaped and compacted to form a wetland using surface water.

The excavator is being used to dig test holes for a wetland and stream restoration project. Deep testing is important when planning projects near streams and rivers, as it can help determine the depth of fill in wetlands, thickness of gravel and clay layers, elevation of groundwater, and depth to bedrock.

runoff should be directed into these wetlands to offset waters that will seep into the ground.

Looking for Buried Wetlands

A test hole can help determine if a wetland lies buried on a site. People have been filling wetlands with topsoil, brush, rock, sawdust, and other waste for hundreds of years, and then covering the fill with soil so the land can be used. Wetlands can also fill from erosion occurring on surrounding lands. Identifying a filled wetland requires the use of one's eyes as well as one's hands. Finding one or more of the following provides strong indications that fill has been placed in a wetland:

- Asphalt, cement pieces, and rocks that seem out of place
- Wood chips, branches, logs, or roots not attached to a tree
- Thick topsoil layer
- Broken glass, soft drink cans, and other trash
- Topsoil beneath mineral soil
- Leaves buried beneath topsoil or mineral soil
- Mixed and variable soil textures
- Patch of rock surrounded by soils with defined layers
- Buried layer of gray-colored soils

Whereas a hand-held posthole digger can reveal wetlands covered by shallow fill, a backhoe or excavator should be used where deeper fill is suspected.

Under ideal conditions, it is best to remove the historically deposited fill in order to restore the wetland. Removing the fill will return the original shape and depth, and may expose seeds covered for hundreds of years that could germinate once exposed to sun and water. However, the reality is that the cost of removing construction fill is often too great for wetland projects. Because it is not possible to compact construction fill so that it will hold water, synthetic liners are often used to build wetlands over filled wetlands.

Damming Streams

For years, wetlands have been created by damming perennial streams at narrow gaps within large, level floodplains. These dams were carefully constructed to withstand the strong erosion force caused by flowing water, with large capacity steel and cement water control structures being used to move water through the dam. In addition, wide emergency spillways were built to carry floodwaters around these structures. The spillways were constructed on gradual slopes near the base of a hill, and were often reinforced with rock and steel to prevent washing.

There are two main reasons why blocking perennial streams is so attractive: creating a wetland with a large

Key to Wetland Construction Techniques

(A) Water seeps into the test hole from underground.......... **Groundwater Technique**
(A) No water seeps into the test hole.......... B

(B) Soils high in clay, thin ribbon sections 1.5 inches or longer*.......... C
(B) Soils not high in clay.......... **Liner Technique**

(C) Soil ribbons are 2 inches or longer.......... **Surface Water Technique (small watershed acceptable)**
(C) Soil ribbons are from 1.5- to 2-inches long.......... **Surface Water Technique (large watershed needed)**

*Length of thin ribbons that break off when formed between thumb and forefinger using a walnut-sized sample of soil after adding water to obtain suitable moisture surface area, and being able to use permeable soils on site, saving the expense of trucking in soils that are high in clay. While dams constructed of permeable soils can be expected to leak, water losses under and through these dams are generally offset by large quantities of water flowing into the wetland from the perennial stream.

Blocking a perennial stream to build a wetland has consequences that should be carefully considered before proceeding:

- Large acreages of deep water will be impounded that are of little value to waterfowl, shorebirds, amphibians, and reptiles.
- The project will be expensive to build and maintain.
- The dam will create a barrier for fish, often necessitating the installation of an expensive fish ladder, which will have to be maintained, to allow aquatic organism passage.
- The wetland will provide habitat for fish, high populations of which will conflict with management for waterfowl, amphibians, invertebrates, and plants.
- Large, high, expensive dams that require much rock and cement must be built for safety, because failure can result in the flooding of land downstream.
- The wetland will be used by anglers that unintentionally disturb nesting and feeding waterfowl, and the anglers' interests can eventually dominate how the wetland is managed.
- Water control structures require regular maintenance in order to prevent blockage by beaver and flowing debris.
- There can be extensive permit requirements.

Sometimes channels are dug to divert waters from rivers and streams into newly-constructed wetlands. This practice is unnecessary, providing the wetland is built to contain either surface water within a basin of clay soils, or groundwater on a site with a high water table. If there is an interest in connecting the wetland to a stream, it is generally less expensive and more acceptable to place the wetland in a floodplain so that it is periodically inundated by waters flowing across the floodplain.

Pumps and Wells

A number of large wetland projects have been built with wells and pumps. These are used to fill wetlands during fall migration to increase waterfowl hunting opportunities, or to remove water in the spring to create conditions for moist soil management. While pumping can be an effective means of adding water to wetlands on demand, the installation, operation, and maintenance of these systems can add tens of thousands of dollars to a wetland project. In addition, impacts to groundwater and the long-term use of fossil fuels are environmental factors worth considering before installing these systems.

Pumps are often necessary to maintain wetlands that would go dry without the continued addition of water to offset losses caused by still-functional diversion ditches, buried drain lines, and subsurface permeable layers. In most situations, money and labor can be saved by rebuilding these wetlands so that they hold water naturally, rather than by maintaining their hydrology with pumps.

Do Your Homework

Private landowner Alleyn Unversaw's experience in building wetlands in partnership with the government illustrates the importance of understanding wetland restoration techniques rather than relying solely on agency personnel to do the right thing.

Alleyn and his business partner Frank Cracchiolo searched for months for a piece of land that they could purchase for hunting, and on which they could build wetlands. Alleyn did his homework and learned how important clay soils are for building successful wetlands. Acquiring soil maps and information from the NRCS, they searched the countryside for wet spots in the fields on farms listed for sale, eventually buying 60 acres of flat ground near Sterling Run Creek in Brown County, Ohio.

After the required waiting period for new purchases, they enrolled the lands in the Wetland Reserve Program (WRP), and imagined soon that there would be wetlands teeming with wildlife. But unfortunately, troubles would soon arise. "The engineer assigned to us had never worked on a wetland project before," said Alleyn. Regrettably, this set the stage for a relationship that would be clouded by stress and failed expectations.

The engineer designed a project they were unhappy with that involved creating only two wetlands, one to be one acre in size and one to be three acres big, leaving the majority of the property dry. At one point while discussing the project, Alleyn asked that topsoil be saved for spreading back in the finished wetland, and was told that would be a waste of money. The engineer stated that no vegetation would grow in the wetland other than cattails, so topsoil was not needed. Alleyn then asked that water control structures be installed, and when informed that they were not allowed, threatened to pull out of the program. A structure was finally approved. The larger wetland received the sanctioned water control structure, while the smaller one did not receive a water control structure of any kind.

The final product was one artificially-looking, square-shaped wetland approximately one acre in size, with a graded, flat bottom and one two-acre wetland having several hundred feet of dam. The small wetland looked like a storm water detention pond with a dam around three-quarters of it and a low-cut spillway. Alleyn stated that the dams never held back much water on either wetland. The two-acre wetland had the potential to be great, but the emergency spillway was cut so low--1.5 feet below the top of the dam--that much of the area the builders thought would be wet was dry.

Being so upset with the results, Alleyn asked that the NRCS Team conduct a review, and was then told the project met contract specifications. "I thought WRP was the best in the nation," he said disappointedly. Over the years, the wetland has been sterile with substantially no wildlife use. "We asked if the spillway on the two-acre wetland could be raised six inches, because it would give us two to three acres more of water," he said. The engineer said that could not be done because of agency regulations. Alleyn and Frank have intensely managed this wetland to encourage plant growth with good results. By partially drawing down water with the control structure, they have encouraged much vegetation growth.

Alleyn shows kindness when talking about the engineer with whom they worked. "He was given a job he did not know how to do, and had all these requirements to follow, but we could have gotten ten times the wetland for half of the cost."

In 2008, Alleyn worked with a different agency, the U.S. Fish and Wildlife Service, to build two additional wetlands on the same piece of land. "Sergio Pierluissi, the private lands habitat coordinator, was wonderful to work with," he said. Agreeing on the design, they built naturally-appearing, shallow water wetlands of varying depths, with topsoil in the bottom. "You wouldn't believe the waterfowl use on these new wetlands"-Alleyn says, "blue-winged teal, shovelers, mallards, and wood ducks." And at night, the music from the frog calls is deafening.

This natural ephemeral wetland is located on a ridge in Menifee County, Kentucky.

This ephemeral wetland was established on a ridge by using a dozer two years ago in Rowan County, Kentucky.

The Construction of Surface Water Wetlands

Chapter 4

Wetlands designed to hold rainfall and runoff are shallow depressions consisting of compacted soils that are high in clay. Water entering these wetlands generally does not soak into the ground. These are called surface water wetlands.

Soils that contain adequate amounts of clay to build a surface water wetland can be shaped and compacted. Those soils that feel gritty, and whose thin ribbons break off in sections less than two inches long are generally unsuitable for use in a surface water wetland.

Soil texture classes are determined by the percent of sand, silt, and clay they contain. Sand particles are the largest; silt is in-between, and clay the smallest. Soils that contain high amounts of clay and/or silt can be packed in place to hold surface water. The following soil textures can be used to form a wetland that will hold runoff:

- Silty-clay
- Silty-clay-loam
- Sandy-clay
- Sandy-clay-loam
- Clay loam
- Clay

"Sand resists compaction and will reduce your ability to form a ribbon or pliable long worm, whereas silt can be compacted," says Dr. David Scott, soil scientist from Okanagan University College. "This means you'll need at least 27 percent clay if there is less than 50 percent sand, and at least 35 percent clay if there is more than 50 percent sand."[28]

Another test is to construct a miniature wetland from the soil. Collect a bucket of the soil, add some water, and then use one's hands and fist to pack a depression in the top layer in the bucket. Fill the depression with a glass of water. If the water is still standing the next morning, one should be able to create a much larger wetland from the soil.

Not being able to dig a test hole because of too much rock indicates that soils are quite permeable and not suited for building a surface water wetland. Seeing rock lying on the surface indicates that more is buried beneath, like the tip of an iceberg. Fortunately, one can often find areas that have less rock by simply moving to higher or lower ground.

Finding coarse-textured soils containing sand and gravel without the presence of groundwater near the surface limits opportunities for wetland construction. One should switch locations or plan to use a synthetic liner to build a wetland on these locations.

Because natural wetlands are often supplied by both groundwater and surface water, one may be tempted to plan a restoration that uses both. Unfortunately, the majority of attempts to blend these two strategies fail. There may be too many unknowns that could affect success, including the presence of buried drainage structures, open ditches, variable soil textures, and burrowing crayfish to accurately calculate a water budget for the wetland. Thus, it is best to plan on building either a surface or a groundwater wetland. However, if the surface water wetland is supplemented by groundwater, or vice versa, the elements of nature are truly working together.

Soils that are high in clay may occur anywhere on the landscape. While clays are most often found on ridges above rivers and streams, large expanses of silt-clay soils can be found across large flood plains. In river valleys dominated by coarse-textured soils, clay can sometimes be found near the base of a hill opposite a ridge. Sometimes clay inclusions large enough to build a wetland can be found by simply changing elevations and checking on either higher or lower ground. Soil survey maps prepared by the NRCS can help point in the right direction.

Unfortunately, it is entirely possible to build a wetland that does not hold water as planned, even when working on top of a historic wetland, because not enough attention was given to determining whether the soils had enough silt and clay to be shaped and compacted to hold water. Wetlands built from soils high in clay can still fail because of improper construction techniques.[29]

Designing the Wetland

Watershed Size

The amount of runoff needed to maintain a wetland can be a concern to individuals planning their first wetland,[30] and the calculations used to determine an optimum watershed size can become so complicated that one may question if it is worth even doing the math. Fortunately, one does not need to calculate a water budget to build a successful wetland. The most important factor to consider is providing a suitable path for water to leave the wetland without causing damage.

East of the Mississippi River, a good guide to follow in selecting the maximum size of a watershed for a wetland is that it should not exceed three acres for every acre-foot of water to be held in the completed wetland. For example, it would be acceptable if the waters flowing off six acres of land were to drain into a wetland one acre in size and two feet deep. Conditions in the field are never perfect, which means that judgment and common sense must be used to decide how much water should be allowed to enter each wetland. When working west of the Mississippi River, the ratio of watershed to wetland size should be greater than 3:1 to ensure the wetland fully develops its planned hydrology.

Do not direct the waters from a perennial stream into a wetland because of the high cost associated with creating a hardened spillway strong enough to withstand the flow of water that will leave the wetland under flood conditions. Costs of time and money are also high for maintaining a wetland supplied by a perennial stream. Frequent inspections, maintenance of water control structures, emergency spillways, and dams are also necessary before minor damage can cause more serious problems.

There are often ephemeral streams near wetland restoration and creation projects that can be directed to enter the wetlands being built. In areas where agricultural activities occurred, these small streams have generally been changed into straight channels that form borders to the now grown up forest being restored into wetlands. A conservative approach to water management involves creating wetlands between the ditches so that the streams do not enter the wetland. These seasonal streams can later be directed to flow down into the wetland should additional runoff be needed to maintain water levels in the new wetland.

In areas where rainfall exceeds evaporation, a wetland that does not have a watershed can be successfully built, as long as soils are high in clay, and can be compacted to prevent water from seeping into the ground. Many natural and restored emergent, ephemeral, shrub, and wet-meadow wetlands can be found on locations east of the Mississippi River that receive little, if any, runoff, including sites on top of mountains.

Landscaping with Wetlands

It takes just as much hard work to build a wetland that looks artificial as it does to build one that looks natural. Wetlands built in the shape of a rectangle or square will stand out for hundreds of years, while those built with natural features will look like they have been there forever. Remember that the most important factor to consider when building a naturally-appearing wetland is the avoidance of straight lines.

When designing a wetland, it is a good idea to seek out and examine natural wetlands in the same area. These can give an idea of the size, shape, and depth of wetlands one may want to create. However, in many parts of the nation, this is not possible because of how extensive drainage actions have been.

Wetlands have signature shapes that one may wish to copy. Prairie potholes in the Midwest, playas in the South, and pocosins along the South Atlantic are typically round. Oxbow wetlands near rivers and streams are narrow, and crescent-shaped. Beaver ponds look like teardrops from above, whereas wet-meadows curve down a hillside like ribbons.

This is a six-month-old ephemeral wetland constructed from clay soils on a mountain ridge within a transmission line ROW on the Daniel Boone National Forest in Kentucky.

Marking the Wetland Construction Site

There are a number of reasons to mark in advance where a wetland will be built. First of all, the markings will help to find the site later, even if it has been recorded on a map or in a GPS unit. Secondly, the markings can help other people, such as botanists, archeologists, and utility workers who may need to survey the site before work begins, find the location if the planner is not present. Having the area marked can also help prospective bidders when contracting, and is important to the equipment operator who will eventually build the wetland.

Hanging brightly-colored flagging from trees and shrubs works well to mark wetland construction sites in forested areas, whereas tall wire flags work best in grassland. It is especially important to mark the perimeter of the area where vegetation will be removed by heavy equipment. Flagging can also form a boundary that equipment operators can be instructed not to cross for protecting neighboring streams, vegetation, wetlands, and archeological sites.

As a guide, an area with a diameter of approximately 80 feet is needed as a minimum when building a small wetland with heavy equipment. Larger clearings are needed when building deeper wetlands to allow room for gradual slopes on higher dams. In forested areas, it is often necessary to remove trees to create a wetland. Looking for small openings between the trees and where trees are less than 12 inches in diameter can save clearing costs. One should look for low stocked canopy gaps in the forest that contain trees that are smaller in diameter that have to be removed.

When building a small wetland on sloped ground, it is necessary to clear an area at least twice as large as the planned surface area of water in the new wetland. This provides room for piling debris, vegetation, and topsoil. Marking an area that is too small for construction does not allow enough room for a natural appearance, and ends in wetlands that have steep slopes and a narrow dam. However, clearing an area that is too large may result in unnecessary costs, and negative impacts to trees and shrubs.

In forested areas, hang florescent pink- or orange-colored flagging on trees and shrubs around the outer edge of the area where trees and vegetation will be removed. When looking at the area that has been marked as being available for building the wetland, it is easy to believe that the finished wetland will flood the entire area to be cleared. Remember that it takes room to pile vegetation, to temporarily store topsoil, and to create gradual slopes on the dam. When examining an area marked for wetland restoration, one can predict that the surface of water in the finished wetland will cover less than one-half of the area to be cleared.

Expect to remove some trees when building a wetland in a forested area. Heavy equipment is generally used to push over trees. These trees do not have to go to waste, as they can be placed back into the completed wetland for wildlife habitat and around the wetland to prevent OHV damage. Felling standing trees on the construction site is an efficient operation with heavy equipment. An excavator or a dozer can remove a clump of 6- to 18-inch diameter trees in a matter of minutes. It is best to leave trees to be removed standing rather than cutting them by chainsaw prior to having the heavy equipment working at the site. Equipment operators can very effectively use the top of standing trees as levers to quickly push over and dislodge even large root masses. Conversely, freshly-cut stumps are much tougher to remove, requiring considerable time and digging in order to oust root masses. Be especially careful if, however, standing dead trees are left on the site, as they can be a safety hazard to people on the ground if they are struck by heavy equipment during construction.

Large Fields

The task of marking wetland construction locations in large fields can appear overwhelming, especially when areas have grown up to shrubs full of thorns. The job can be made easier by mowing with a tractor and bush-hog. This will increase the visibility of drainage ditches, depressions, and slight rises, which will aid in deciding where wetlands should be located. Mowing can also reduce the cost of an *Archeological Survey*, and provide a more inviting area for prospective bidders to examine.

When building a large wetland, it is a good idea to mark an area that is large enough to build the dam, obtain soil for the dam, and to pile vegetation and topsoil. This generally involves marking a strip from 200 to 250 feet wide along the entire length of the dam.

Designing the Dam

Most wetlands are formed by constructing a dam that prevents water from flowing downhill to a stream or river. While a dam is not needed to form a wetland on level ground, there are few areas flat enough where one can be built without a dam.

Waters in natural wetlands are often maintained by a dam. Natural dams are typically formed of clay, and generally blend into the surrounding landscape because they have gentle slopes with trees growing on them.

The shape of the dam will affect how natural the wetland will appear. Unfortunately, many wetlands are built with high dams and steep slopes to reduce construction costs and to create deep water. Generally, the more gradual the

Gradual slopes were placed on the front and back of this newly-constructed dam for an emergent wetland in Fleming County, Kentucky.

slopes and the wider the top of the dam, the better it will blend with its surroundings. Other advantages include reduced maintenance costs associated with muskrats and beaver tunneling, less potential for flood damage, fewer leaks from tree roots, and greater plant diversity. OHVs are also less likely to damage a wide dam with gradual slopes.

Many people who are interested in wetland restoration are strongly opposed to constructing a dam to bring back the wetland. They fail to realize that the majority of natural wetlands are maintained by a dam, and that these natural dams are generally wide with gradual slopes. Therefore, in order to restore wetlands that look natural, and gain greater acceptance of these recreated environments, it is best to place slopes of five percent or more gradual on both the front and back of the dam, and shape the top of the dam so that it is 12 feet or more wide.

Determining the Location and Height of the Dam

To determine how high the dam ought to be, it is necessary to measure the slope on the area where the wetland is to be restored. The height of the dam should be equal to the difference in elevation between the highest desired water level and the lower edge of the work area. It is best to first clear the work area of trees and shrubs before marking where the dam will be built.

It may be helpful to think of a bowl of milk when contemplating where to build the dam. When the bowl is set on a flat surface, the space between the milk and the top edge of the bowl is even all around, and there is no need for a dam. However, when the bowl is tipped, like the slope on the surface of the ground, milk will spill out of the bowl unless the lower edge has a rim to keep it in. The dam on a wetland is that rim on a bowl that keeps water in the wetland. The most efficient way to determine how high the dam should be is to use a simple construction level on a tripod and a rod to measure slope over the proposed work area. Unless there is a laser level available, two people will be needed for this job. The level should be set up near the upper edge of the area, overlooking the possible construction site. Ask the worker with the survey rod to stand along the lower edge of the area and take a reading with the level; then ask him to move to the upper edge where the water should peak for a second reading. The height of the dam needed to flood the area is equal to the difference between these two elevations. However, since a spillway is necessary for handling overflow waters, and the spillway is usually made 12 inches below the top of the dam, one should add 12 inches to the difference between the upper and lower elevation for the maximum height of the dam.

A simple construction level can be used to determine slope on a wetland construction site. (Barry Nichols photo).

Time spent determining elevation differences on a site is time well spent. It will give a good sense of how large a wetland will fit onto the site, and will reveal that areas that appear to be flat from every standpoint usually are not. Keep in mind that the dam can be irregular, and that sometimes, following the same contour as the land can provide a larger, more naturally-looking pool.

When marking the work area for a wetland restoration project, one typically flags an area of land that has less than a three-foot change in elevation from the upper to lower edge, and is bordered by features not to be disturbed, such as large trees, steep ground, streams, rivers, trails, or cultural resource sites.

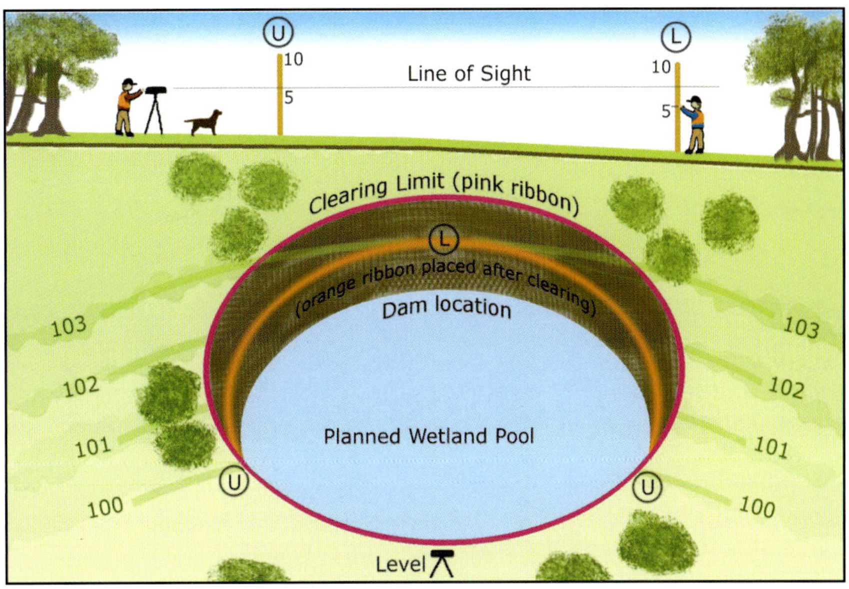

Marking the wetland construction site

Here is an example of how to mark the work area for a typical restoration site. Imagine that one is marking a circle that represents the approximate shape of the area to be flooded. Start by standing with the survey rod along the upper edge of the proposed work area where one would like water to reach in the new wetland. In this instance, the number that is read on the rod is 5.0 feet. Tie a ribbon at this location. Now walk downhill with the rod until the number read is 8.0 feet, or three feet lower than the first reading. Tie a flag at this location ("L" on drawing). This flag marks where the deepest water will occur and highest portion of the dam will be built.

Now mark where the dam will begin and end, which is essentially the same as the left and right upper sides of a circle. Walk to the left side of the worksite and place a flag where the reading on the rod is the same as the one taken along the upper edge, or 5.0 in this example ("U" on drawing). Walk over to the right side of the area and tie a flag where the reading is also 5.0. Now, flag an arc between these points along the downhill edge to represent where the center of the dam will be built. The arc begins on one side where the reading is 5.0, goes downhill to the lower flag hung at 8.0, and then uphill to the other side where the reading was also 5.0. When finished, the lower edge of a circle will have been marked that shows where the dam is to be built.

The flagging tied to mark the middle of the dam to be built may need to be shifted uphill, so that there is room between these flags and the lower edge of the work area for piling vegetation, topsoil, and for forming the back slope of the dam. The path of the flagging between the beginning and the end of the dam can also be adjusted to give the wetland a more natural shape.

While the construction level is in place, it is a good idea to drive a nail into a tree that is equal to the elevation of the top of the dam to be built. In this case, a tree would be chosen along the lower edge of the work area, but not inside it. Also drive a nail into a tree at the same elevation as the top of the dam to be built. If a tree is not available, a stake can be driven into the ground so that the top of the stake is equal to the top of the dam. This reference mark will be most helpful when it comes to checking the height of the dam during construction. On the day of construction, one simply returns to the site, sets up the level in an out-of-the-way spot with good visibility, places the survey rod on top of the nail, and takes a reading with the level. The number read on the rod will be the same elevation as the top of the dam to be built.

These examples show how to determine the height of the dam on areas on different slopes:

Steep hillside location

Lower edge rod reading = 8.0
Upper edge rod reading = 5.0
Difference = 3.0
Add for spillway = 1.0
Maximum height of dam= 4.0 feet
Rod reading equal to the top of the dam at time of set up = 4.0
Rod reading equal to water level and spillway at time of set up = 5.0

Sloped location

Lower edge rod reading = 8.0
Upper edge rod reading = 6.0
Difference = 2.0
Add for spillway = 1.0
Maximum height of dam= 3.0 feet
Rod reading equal to the top of the dam = 5.0
Rod reading equal to water level and spillway at time of set up = 6.0

Flat location

Lower edge rod reading = 6.0
Upper edge rod reading = 5.0
Difference = 1.0
Add for spillway = 0.5
Maximum height of dam= 1.5 feet
Rod reading equal to the top of the dam = 4.5
Rod reading equal to water level and spillway at time of set up = 5.0

The height of the dam is basically equal to the difference in elevation between the upper and lower edge of the area to be flooded.

Should more than a three-foot change in elevation between the upper and lower edge of the work area be found, consider moving the lower edge of the dam uphill so a dam that is higher than four feet will not be required. Another option on steep slopes is to divide the site into separate wetlands, each with a dam less than three feet high. Wetlands with low dams and shallow depth contain a greater diversity of plants and animals than deep wetlands with high dams. Low dams are also much cheaper to build and maintain than higher dams.

The following chart shows the amount of distance that should be left between the lower edge of the work area, or the clearing limit, and the florescent flags that mark the center of the dam for various dam heights to be constructed:

three-foot high dam = 40 feet
two-foot high dam = 30 feet
one-foot high dam = 20 feet

In wooded areas, different colored flags may be used below the dam to mark the edge of the clearing limits for the contractor to use as a guide during construction. It takes practice to learn where the dam should be located. In order to keep wetlands shallow, a dam should be constructed for every three-foot change or less in elevation over the work area. Another way to picture this is to think of a topographic map where a dam is placed on every three-foot contour interval.

When designing a wetland in an open field, the tendency is to try to inundate the entire opening by constructing one high dam. Unfortunately, these wetlands can become exceedingly deep, and consequently, be of little value to plants and animals. Building one high dam is generally more expensive than building several low dams. There are other advantages to building a series of low dams instead of one high dam, which include creating more naturally-appearing wetlands, having the ability to manage a series of wetlands at various water depths, and reduced maintenance costs due to muskrat damage.

Using a Construction Level

The construction level is an indispensable tool for building a wetland. Just as those who drained wetlands hundreds of years ago used a level to dig ditches to destroy the wetlands, workers are now using it to bring them back. A level is a plain telescope that is focused with a knob. It contains a crosshair like a riflescope. The level is mounted on a tripod, which is set up along the upper edge of the worksite so one can view the potential wetland area. A survey rod is always used with the level. The rod is the same as a rigid tape measure, and can be marked in inches and feet, tenths of a foot, or meters and centimeters. By reading the numbers on the rod, how much slope is present over an area can be determined.

A level is easy to use, and one does not have to master trigonometry to become proficient in its use, as there are no angles involved. Even grade school students can be taught how to use a level to build a wetland at their own school. The level will help one see where water will flow into and out of the wetland before it is built. It will help determine the height of the dam needed, and whether the top of the dam built is level. Learning how to use a level can save costs associated with having an engineer design and layout the wetland project. Fortunately, heavy equipment operators almost always have a level on the jobsite, and are often willing to show the landowner or volunteer how it is used.

Not Flooding the Neighbors

Use the level and rod to measure the elevation of roads and neighboring property to make sure that the dam designed will not flood these lands and improvements. Often the elevation of the dam is controlled by those areas one must protect. Landowners can become very disturbed when they see water backed up to the edge of their property. Sometimes wetlands builders and restorers have had to dig diversion ditches and pile soils along property lines to ensure that neighboring lands were not affected by surface and groundwaters caused by their initial efforts. Keep the maximum height of the dam below the elevation of neighboring property so that if beaver block the water control structure and spillway, other ownerships and improvements will not be flooded.

> *One may be able to borrow a construction level from a contractor or local government office to use on a wetland project. Laser levels are nice, but a great job can be done with the same simple level that has been used to drain wetlands since the 1800s.*

A Question of Contour Maps

One does not need to prepare a contour map to successfully design and build a wetland. USGS topographic maps generally have the detail needed to find sites level enough to mark a wetland. While it is easy to become consumed by the process of preparing a detailed survey map, one's time is better spent walking around an area to get a better sense of where wetlands should be built.

With a little practice, one can spot where wetlands can be built on a contour map. In mountainous areas, this will usually be how the more level places are found. Identifying these areas in advance and circling them for examination will make the time spent outdoors looking for wetland construction sites more productive. After a while, the level benches on hillsides, canopy gaps in the woods, and wide ridge-tops will begin to stand out as excellent locations for building wetlands.

Having a map available with one-foot contour intervals can be a valuable aid when designing a complex of wetlands in a large open field. However, to be effective, there must be a number of points marked on the ground that correspond to the map so that one knows where one is on the map to mark dams for the wetlands. Using GPS equipment mounted on an OHV will speed the preparation of survey maps of fields larger than five acres.

It may be a waste of time working to complete a detailed contour map of large fields that have grown up to trees and shrubs. Trying to layout wetlands using a contour map in a dense growth of trees and shrubs can be almost impossible because of the difficulty in seeing features like ditches and slope, and not knowing for sure where one is on the map. Because these areas are difficult to survey, it is better to first clear the lower portion of each field's perimeter with a dozer or tractor and bush-hog so that it can better be determined where the dams should be located.

Designing in a Floodplain

Building wetlands in a floodplain can be beneficial because so many wetlands historically occurred along rivers and streams. However, special precautions must be taken to design and construct these wetlands so that they will not cause erosion or flood neighboring lands. Because permits are usually required for building a wetland in a floodplain, it is important to find out if the proposed project is located on one before beginning. Here are some ways to identify a floodplain:

- The landowner or a neighbor says the area floods.
- Large logs are present in fields.
- Vegetation is matted down and points in one direction.
- Vegetation and trash are wrapped around the base of trees.

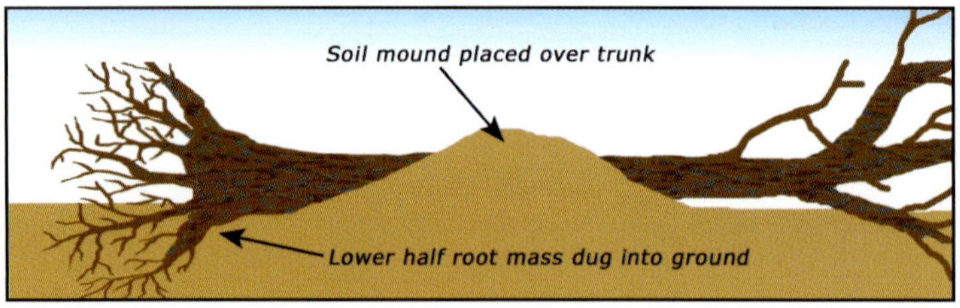

A method of anchoring logs in a floodplain.

- Grasses are trapped by fences and are hanging from trees and shrubs
- The area is named as a floodplain on government or insurance company maps

Some communities require that a *Flood Assessment* be completed before actions such as dam construction are allowed to take place on a floodplain. The main purpose of this is identifying the risks of the proposed action on neighboring property and improvements. A *Flood Assessment* can cost over $10,000 to complete due to the need for finishing a detailed survey and analysis. Design factors can be adopted to eliminate the demand for conducting a *Flood Assessment* by lessening the risk of changing flood flow by the wetland project. These factors include:

- Not building high dams
- Not building dams with steep slopes
- Not building dams perpendicular to flood water flow
- Lowering the profile of the wetland by doing more excavation and less dam building
- Spreading soils removed in low places down slope from the dam being built, or in low ridges paralleling the flow of floodwaters
- Moving excavated soils to higher ground above the floodplain

Fish can be expected to colonize wetlands constructed in areas that flood. Avoid problems with unwanted fish by constructing wetlands so that they are shallow enough to dry, or by installing a drainpipe.

Anchor logs and trees placed in wetlands near rivers and streams so they will not wash downstream in a flood event. A simple way to do this is to place a pile of soil over a portion of the tree trunk. Placing larger trees with soils still attached to roots will accomplish the same purpose.

This small ephemeral wetland was constructed within the floodplain of a creek. No dam is visible on this wetland, which is supplied with groundwater.

Heavy Equipment

Key machines for building wetlands are the dozer and excavator. Both perform best when mounted on tracks instead of tires. Heavy equipment mounted on tracks is much less likely to become mired, and to cause rutting than machines on tires. The dozer is essential for scraping, moving, shaping, packing, and spreading soils, whereas the excavator excels at digging, sorting soils, and placing wood and rock. The dozer should be a model that has a seven- or eight-way tilt blade designed for shaping soil, rather than using a huge machine with a rigid blade made for pushing dirt as on mining jobs.

A dozer will suffice when restoring small wetlands on ridgetops far above streams and rivers, but both an excavator and a dozer are needed when working near rivers and streams. The excavator is used to dig below the bottom of crayfish holes and drain lines that may be present, and is the only machine suitable for removing buried layers of permeable soils often found near flowing water.

The various types of heavy equipment have their advantages and disadvantages for wetland construction. The photos show the various pieces of equipment as they appear and are being used on the job:

Excavator:

- Toppling trees on the worksite
- Digging ditches to drain a worksite prior to construction
- Removing and sorting gravel and clay layers for groundwater dam construction
- Removing buried drainage structures
- Planting large snags and trees
- Creating wetlands that use a synthetic liner
- Restoring groundwater wetlands in saturated soils
- Loading trucks with soil and rock

Small excavators with a push blade work well for building small wetlands requiring liners at schools where there is limited access.

Alex Hunley (shown here) displays one of hundreds of clay drain tiles removed by an Excavator (100-Series) restoring a forested wetland.

Dozer:

- Moving toppled trees
- Scraping and moving topsoil
- Filling and packing the groundwater dam
- Building, shaping, and packing the dam
- Shaping the wetland pool area
- Replacing topsoil in the completed wetland.

A medium-sized dozer (105 HP) removes topsoil used to fill the wetland.

This small dozer (71 HP) removes topsoil from a wetland construction site. (Lin Whitley photo)

Backhoe:

- Repairing muskrat and beaver damage caused to dams
- Moving small quantities of soil

This backhoe (2WD) is being used to dig test holes.

A track-steer is packing the depression for a small wetland being built with a synthetic liner.

Track-steer:

- Constructing small wetlands using synthetic liners
- Moving small quantities of soil

A track-steer on wheels carves a depression for a small wetland being built at a school using a synthetic liner.

Scraper:

- Moving large quantities of soil over long distances.
- Compacting dams
- Leveling large areas

A scraper is being used to lower elevations on a large wetland project.

This tractor, mounted on tracks, can efficiently pull one scraper on wet soils and two scrapers on dry soils.

Contracting

Working with a skilled contractor who cares about what the wetland builder wants to accomplish practically guarantees that the project will be a success. Thus, it is well worth spending the time to find an experienced contractor in the area who is willing to work with one.

Small businesses can offer much to the wetland builder. Often family-owned, these companies may have lower overhead costs so they can work for a lower price than a larger business. Since small companies are generally local, and their survival depends on repeat business, they are usually willing to go the extra few yards, or acre, or even mile, so that the customer is pleased with their work. The ideal situation is to have a group of contractors who are trained in wetland construction available for work in one's area. This is best accomplished over time by having wetland construction jobs available each year, and awarding these jobs to area-based businesses.

Paying a contractor by the hour for each piece of equipment that is operating on the worksite can be a very effective use of one's funds. This method often results in the lowest price because the contractor knows that payment will be received for the exact work required on the site. The contractor does not have to inflate the bid to cover possible unknowns, such as buried gravel layers and wet soils that might increase the amount of time needed to complete the job. Paying by the hour allows both the contractor and the wetland builder the flexibility to modify the design of a wetland as needed to respond to changes in soil texture and moisture without having to negotiate and pay inflated prices for contract modifications.

When planning to build a number of wetlands, it is wise to prepare a written contract for the work. The contract contains specifications for the job that are indisputable in comparison to verbal instructions. The contract makes it possible to hold fair competitive bidding, which can result in a lower price for the wetland project. It is essential to have a written contract signed by both parties when dealing with persons one has not worked with before. In this way, both parties will know from the beginning exactly what is expected, and there will be a legal means of denying payment if the job is not completed as specified.

There are basically two kinds of contracts that can be used to complete wetland and stream restoration projects: a service contract and a construction contract. Under a service contract the contractor is paid by the hour for productive work that is performed by each piece of heavy equipment with an operator. A contractor representative is generally on site directing work on a daily basis. Comparatively, under a construction contract the contractor is paid by the job for building wetlands and streams according to detailed drawings and written specifications. The contractor is generally the one who is directing operations each day.

The costs associated with building wetlands and streams are generally three times more expensive under construction contract than a service contract. The difference can be caused by a number of factors, including the high costs associated with an engineering design and survey, higher wages required by the government, increased charges for changes made during construction, and the contractor pricing the job to account for any unknown conditions that may be encountered.

One consideration that can help keep smaller contractors on the job is to provide them with partial payment as work progresses. When working on a number of wetlands, it is often possible to set up a schedule where payments are submitted for each group of wetlands that the contractor completes. Money can be saved by paying the smaller contractor often. This aids the wetland builder's cash flow, and helps the contractor avoid loans for operating expenses while the job is underway.

Require that the contractor wash equipment before bringing it to work on one's job site to reduce the possible introduction of non-native invasive plants into the new wetland. Dozers and excavators often arrive at the worksite with tons of soil wedged between the tracks. There is no way of knowing if this soil contains undesirable weeds, such as Japanese grass, purple loosestrife, reed canary grass, or Johnson grass. Include this stipulation in the signed contract, and check the equipment when it arrives to ensure that it is clean.

Preparing the Site for Construction

The first step to restoring a wetland involves removing trees, shrubs, grasses, and topsoil from the area where the dam will be located, and the area in front of the dam where soil will be removed to build the dam. Save the vegetation and topsoil for later spreading over disturbed soils by placing it in piles along the upper edge of the work area. Leave room behind the piles so the dozer can later push the material downhill over the completed wetland. Construction costs may be reduced by only saving and spreading half of the topsoil. The upper layer, containing a majority of organic material, is saved in piles, while the lower layer, containing more mineral soil, is heaped in a

A pile of woody debris cleared for construction is placed back in a new wetland by a large dozer (130HP) to improve habitat on the Wayne National Forest in Ohio.

row along the lower edge of the work area for use in shaping the backside of the dam.

For large wetlands that involve the construction of long and low dams, vegetation and topsoil should be removed in a strip up to 250 feet wide along the entire length of the dam being built. Trees and shrubs are piled out of the way for later placement back into the wetland. When clearing trees and shrubs, place them in a number of small heaps along the uphill edge of the worksite, where they will be easy for the dozer to push back into the completed wetland, saving problems associated with dissecting larger piles where brush is intertwined.

The upper layer of the topsoil is stripped, moving uphill, and piled along the upper edge of the cleared area. The lower half, containing more mineral soil, is stripped downhill, and piled along the lower edge of the cleared area. The topsoil piled on higher ground will later be spread in the new wetland, and the topsoil placed along the lower edge will be shaped into the backside of the dam.

It is important to remove all topsoil from the dam location and the area in front of the dam where soil will be taken to build the dam. Topsoil contains roots and leaves that will cause leaking if left under the dam. Since the organic material and sand in topsoil cannot be packed, if placed in the dam, there is a good chance the dam will leak. Topsoil is generally darker than mineral soil, and contains visible stems and roots. In forested areas, it is uncommon to find topsoil layers thicker than four inches; however, in agricultural areas, it may be necessary to remove a 12-inch layer of topsoil before building the dam.

A layer of gravel, sand, or silt left under a construction site can drain the new wetland much like the hole in a bottom of a flower pot.

Use staged construction when building larger wetlands that have long dams to reduce erosion and to avoid being stopped by heavy rains. This involves only stripping vegetation and topsoil from a section of dam that can be built in a day. Rain can saturate mineral soils in an area where topsoil was removed to the point where the soil cannot be used to build the dam. These wet areas can take a week to dry out, and cause considerable delay to the project, while areas left undisturbed can generally be cleared the day following a rain, and the soils beneath will be dry enough to use the same day to build that section of dam.

Subsurface Permeable Layers

When building a surface water wetland, it is critical to identify and to make allowances for the presence of buried gravel, sand, or silt for success. These permeable layers can pull water from the wetland just as efficiently as a buried drain line. Subsurface gravel, sand, and silt layers are often found in valleys, and are commonly discovered near rivers and streams. While infrequent, they can also occur on high ground along mountain ridges.

It is best to verify the presence of buried permeable layers prior to construction. Early identification allows for determining the best construction technique, an adequate amount of funding, and suitable heavy equipment for building the wetland. The most reliable way to confirm below-ground permeable layers is to use an excavator to dig one or more deep test holes on a site.

There are three main techniques that can be used to build a wetland on a site containing subsurface layers of gravel, sand, or silt:

- Construct a groundwater dam that is based on an impermeable layer of rock or clay. This is generally the most effective and least expensive technique.
- Create a two-foot thick layer of compacted clay over the entire bottom of the planned wetland.
- Install an aquatic-safe synthetic liner and protect the liner from puncture by using thick geo-textile fabric and soil.

Crayfish and Wetland Success

Considering how and why crayfish dig deep, vertical tunnels into the ground, and planning on that activity occurring at a work area, is critical to the success of wetland restoration across much of North America. Runoff flowing down crayfish burrows into buried permeable layers of soil is one of the most frequent causes of wetland failure.

Crayfish are crustaceans that look much like small lobsters. They have gills to breathe, and are dependent on aquatic habitats for their survival. Found throughout North America, they can survive for a time out of water by breathing moist air. All are capable of excavating burrows that reach down into groundwater to survive periods of drought or freezing weather. Some crayfish burrows contain surface water just like a miniature pond, while others intercept the water table and contain groundwater like a small well. In dry weather, crayfish can remain dormant for months in their burrows, keeping their gills wet for respiration by resting partially in water. Crayfish may leave their burrows at night under conditions of high humidity to feed on plants and invertebrates, or to seek mates.

Crayfish in a constructed burrow that extended 20 feet below the surface on a wetland restoration site in Lincoln County, Kentucky.

Crayfish are grouped into three categories: the primary, secondary, and tertiary burrowers. Primary burrowers live most of their lives in holes they have excavated in the ground, and are basically independent of standing or flowing water. They construct an elaborate system of vertical burrows with multiple chambers, branches, openings, and shafts that extend deep into the ground, allowing them to survive periods of dry weather. The primary burrowers can be found in ephemeral wetlands, wet-meadows, drainage ditches, and even agricultural fields.[31] The holes they dig can be from one to two inches in diameter, and can reach depths of 30 feet. Crayfish burrows can be closely set together, as measured in Lincoln County, Kentucky at a density of 6,000/acre, being only 32 inches apart.[32] Because these burrows often branch and connect, wetland dams constructed by simply piling and packing soils over the surface are certain to fail.

Secondary burrowers also dig deep holes, and can live in them most of the year; however, they may periodically go into surface waters to feed or to look for mates. Their burrows are simpler, and usually have an entrance leading to permanent water, such as a stream or wetland. Tertiary burrowers generally live in the permanent flowing waters of creeks and rivers, and will build shallow, simple, vertical burrows into the water table to survive drought. Crayfish native to eastern North America belong to the family *Cambaridae*, which includes primary, secondary, and tertiary burrowers. Those crayfish native to the west belong to the family *Astacidae*, which are only tertiary burrowers. Primary and secondary burrowers have the greatest potential to impact the success of wetland projects because of the great depth and complexity of their burrows.

One can determine if crayfish are present on a construction site by looking for the open, round entrance holes to their burrows on the ground surface, and for mounds of soil they have pushed out of the ground, which often cover their holes. These mounds are made from many small pellets, and before rains, look like chimneys. Some chimneys are as high as 12 inches. When it is very dry, the chimney may not be apparent, but one can find burrows whose entrances are plugged with soil. These plugged burrows most likely contain a crayfish waiting for wetter weather.[33] Crayfish burrows can still be present even if their holes, chimneys, or mounds are not seen prior to construction, as these features are easily obscured by vegetation. In many cases, crayfish burrows are found only after vegetation and topsoil have been scraped off the surface of a construction site.

The excavator digs below the bottom of crayfish burrows in creating the trench for the groundwater dam. The dark hole is the bottom of a crayfish burrow, which extended into groundwater that can be seen pooling in the trench.

Wetlands can be constructed to perform in concert with crayfish burrowing activity. This is accomplished by taking certain precautions when building the dam. It is critical to collapse, fill, and pack soils over all crayfish burrows found beneath the location of the future dam to prevent waters from flowing under the dam after it is built.

Crayfish burrows act like pipes that allow water to soak deep into the ground, and as conduits for runoff to reach permeable layers beneath the surface. When constructing the groundwater dam, special attention must be given to digging to the bottom of crayfish holes. Collapsing only the upper portion of their burrows is ineffective, as a crayfish left alive in the bottom of its burrow can clear the plugged hole following construction, causing the wetland to fail. It is essential to excavate and pack even old crayfish burrows, because when partially filled with soil and vegetation, they are still capable of carrying surface waters under the dam.

Because of the protection deep burrows provide, crayfish often survive and continue to use areas following human-caused soil disturbance on the surface. During wetland construction, vegetation and soils are typically scraped from the surface over large areas in layers up to three feet deep. Returning the next morning, workers may find crayfish burrows that have been filled and crushed during the previous day's activities were freshly opened by crayfish trapped in them overnight. This behavior may explain how crayfish continue to use burrows following the clearing of fields, construction of subdivisions, golf courses, or wetlands. Crayfish trapped in their burrows simply dig their way out when the ground-disturbing activity ceases on the surface.

It is wise to construct any wetland as if crayfish are present, even when no evidence of their burrowing can be found. Populations of primary and secondary burrowers have been introduced into Arizona and New Mexico, and have become widespread and abundant in the Colorado River Basin where they are not native. In many cases, these populations became established following purposeful releases by agencies for aquatic weed control, and by individuals for food and live bait farming operations.[34] Crayfish are capable of walking long distances over dry ground. Individual northern crayfish were documented traveling 6,060 meters overland from one pond to another in Arizona.[35] Because a newly-restored wetland can provide suitable habitat for crayfish, it is reasonable to assume they could find it someday.[36] It is safest to construct thinking that, if one builds it, they may come.

Crayfish burrowing activity explains why, many times, a wetland constructed by lining the bottom of a basin with clay fails. When wetlands constructed in this way evaporate naturally or are drained, crayfish living in them can be expected to tunnel into the bottom to survive, and in so doing, will puncture the layer of clay and cause the wetland to fail.

Importance of the Groundwater Dam

It is essential to construct a groundwater dam if restoring a wetland designed to capture surface water, or if attempting to raise the elevation of the water table. A groundwater dam is a zone of packed soils high in clay that prevents water from leaving the wetland underground. Natural soils are loose and porous, and if left unpacked under the dam, they form a conduit for waters to leave the wetland. A groundwater dam is needed wherever a dam is built, even if a dam is only one inch high. The dam and groundwater dam are inseparable, and should be built as one unit. The groundwater dam stops water from flowing under the dam, which is the number one cause of wetland failure.

Groundwater Dam Construction

The removal of soils from under the planned dam location, and subsequent replacement and compaction of these soils, is known as constructing the groundwater dam. Construction of a groundwater dam represents a critical step that will almost guarantee the success of a wetland project. Water could flow under the dam via buried drain structures constructed of wood, rock, clay, and plastic, by crayfish burrows, or through subsurface layers of gravel, sand, and topsoil. Crayfish provide the means for surface water to enter buried permeable layers and travel beneath the dam. The groundwater dam raises the water table, causing soils within the wetland to become saturated and hold pools of water.

A groundwater dam is needed to successfully block the flow of intermittent or ephemeral streams when building a wetland. During most of the year, waters flow in these streams below the surface in gravel and sand layers. To prevent waters from leaving a wetland by blocking a stream that flows infrequently, it is necessary to interrupt the subsurface flow with an underground dam.

The depth of the groundwater dam is related to the elevation of the worksite. On high ground, it may only be necessary to dig a trench deep enough to cut through the topsoil, tree roots, and any drainage structures that may be present. On some ridge-tops, where crayfish are absent, it is only necessary to cxcavate 24 inches deep to create the groundwater dam. However, one can expect to dig much deeper when building near a stream or within a 100-year floodplain.

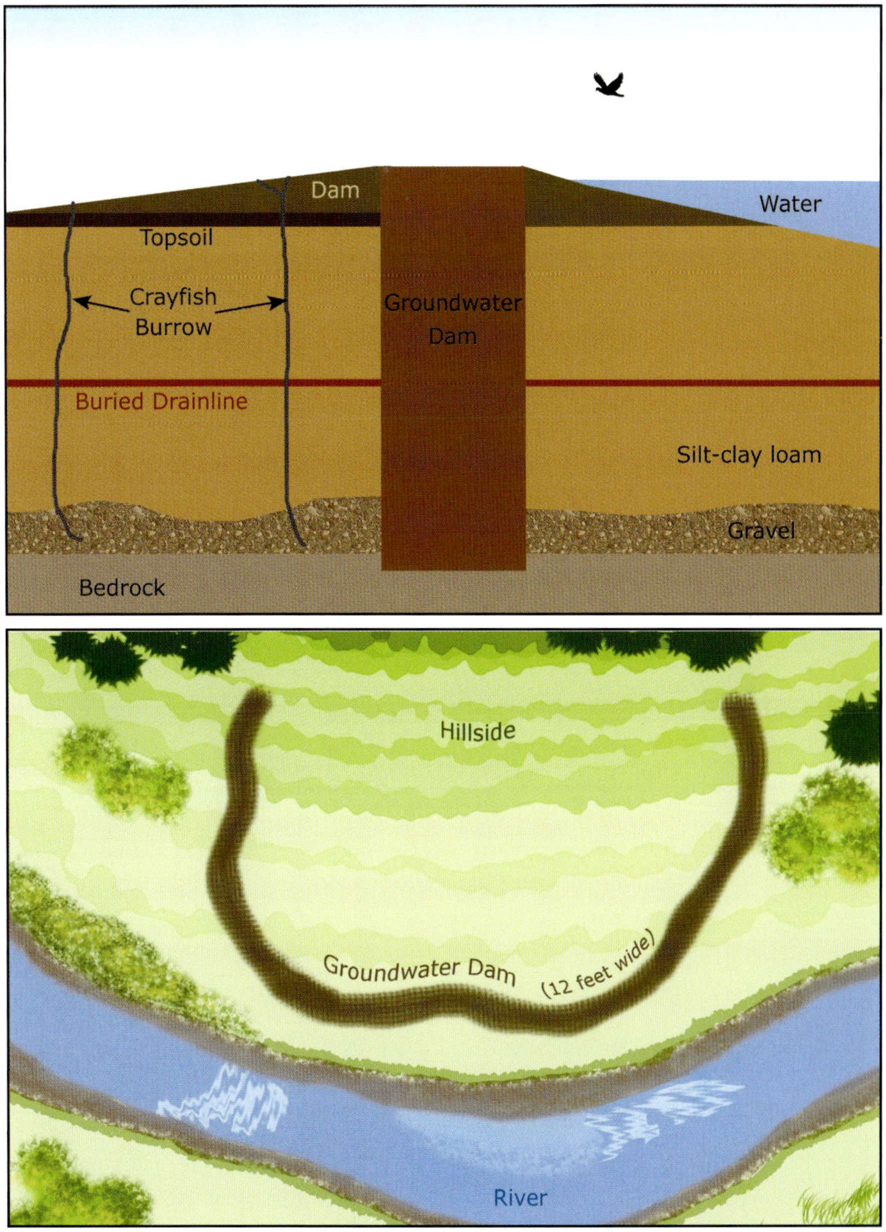

Groundwater dam-Profile view (top)
Groundwater dam-Plan view (bottom)

Evidence of farming in an area should serve as a strong warning that a deep groundwater dam is needed below the dam. Old fields, shallow ditches, broken down fences, and stone foundations all indicate historic agricultural activities. It is important to remember that where signs of agricultural practices are found, people regularly disturbed the ground to make it more productive, and farmers whose living depended on crops they raised likely toiled for generations to remove water from the areas one now wants to flood.

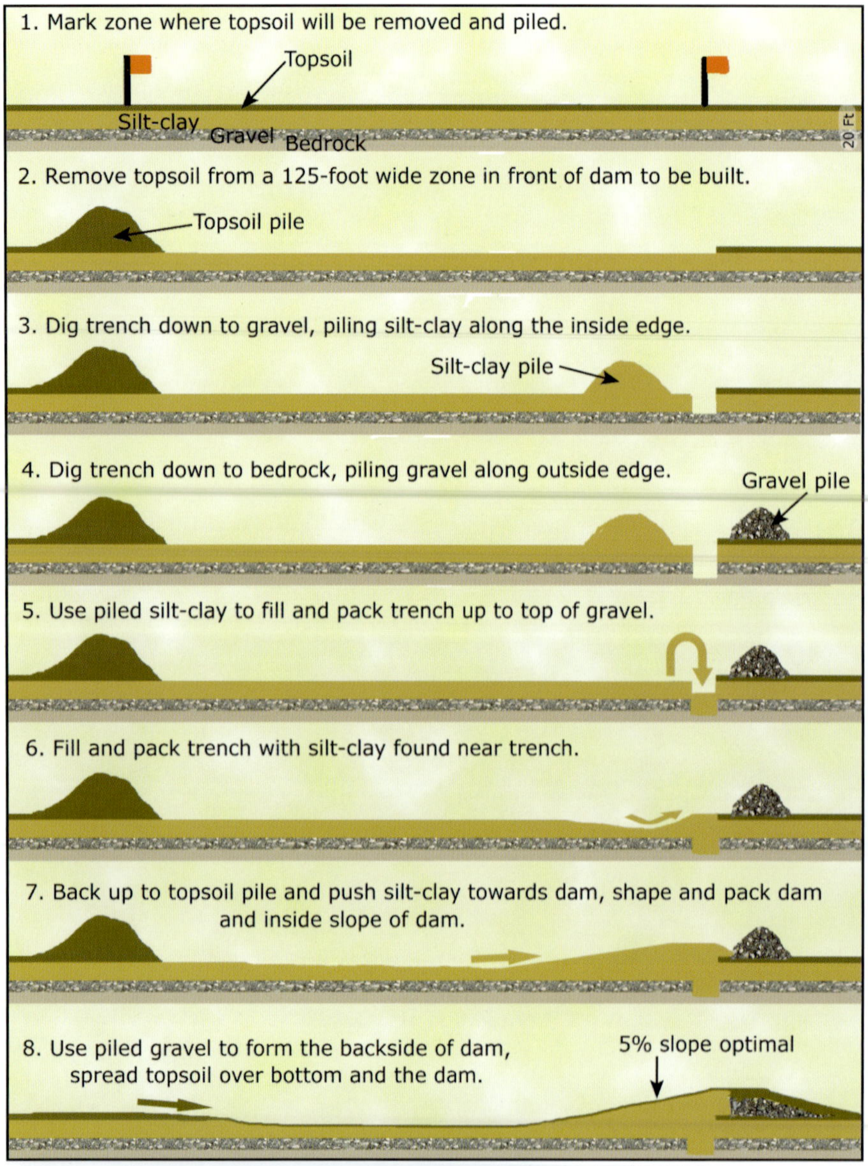

Steps for building a groundwater dam. (top)

Water travels underground in limestone in different layers, resulting in ice hanging on a road cut, making it difficult to build a wetland from clay soils over the top. Cracks in limestone act as the hole in the bottom of a flower pot to drain surface waters.

Some individuals try to use a plow or backhoe to break the path of buried drain lines to create a wetland. Unfortunately, this is not enough. Water pressure can be so great that waters will continue to flow through the soils surrounding broken drain lines. Plugging the end of a buried drain line with concrete is just as ineffective. The pressure within the line causes water to blow out the numerous gaps between the pipes, and it will continue flowing. Outlets for drainage systems covered by six feet or more of soil can also sustain their current uninterrupted. Drain lines maintain their flow even when crushed, as the water pressure creates a new channel that will rejoin with intact drain lines further downstream. Exposing all drain lines where they pass beneath the dam and blocking them with a groundwater dam is the only reliable way of inactivating these systems.

The soil used to build the groundwater dam is generally the same texture as that used to build the dam, which means that it must have enough clay to form a ribbon between one's thumb and forefinger at least two inches long before breaking. In a majority of cases, soils removed when excavating the trench for the groundwater dam that contain an adequate amount of clay are placed back in the trench. This soil is then supplemented with soil excavated from inside the future wetland. Due to the high expense, soil is seldom hauled in from outside the wetland area to fill the trench.

The trench dug for the groundwater dam must be deeper than the bottom of any crayfish burrows present on the site, and where possible, be based on an impermeable layer of rock. Water pooled on the surface of the subsurface rock layer indicates a suitable foundation on which to build the groundwater dam. This layer of bedrock is generally the same as that found at a lower elevation in the bottom of the nearest river or stream. This technique may not work on limestone, as it contains many cracks and sinkholes through which water can flow deep into the ground.

The groundwater dam is built on the same location where the dam for the wetland will be built. One end begins on high ground at the upstream edge of the wetland being built, curves downhill in an arc, and ends on high ground at downstream edge of the same wetland. The groundwater dam will have the shape of a capital "C" from above, with both ends tying into the hillside. The finished elevation of the groundwater dam must be at or above the planned elevation of water in the new wetland.

Begin by removing topsoil from the location of the dam to be built and the area in front of the dam where soil will be removed for building the dam. The width of the area to be stripped can range from 150 to 250 feet wide. The wider zone is used when building a tall dam and on sites where a thick gravel layer is present.

Use florescent flags to mark the area where topsoil will be removed. Take only the topsoil from the length of dam that can be built in the same day, generally 400 feet or less. Removing more will cause problems if it rains, as the area will collect water and become soft, and these saturated soils cannot be compacted to form the groundwater dam.

Topsoil should be stripped with a dozer to a depth where all plant material is removed, along with the majority of roots. It is acceptable to leave in place broken pieces from deeply rooted trees. The presence of lighter colored mineral soils on the surface indicates that the topsoil has been stripped to an adequate depth. Push the upper layer of topsoil, containing organic material, uphill for later spreading over disturbed soils in the bottom of the completed wetland. Shove the lower layer, containing mineral soil, into a long, continuous row along the downhill edge of the cleared area for use in the backside of the dam.

The excavator next digs the trench on higher ground above the wetland being built. The trench should be dug wide enough so that a dozer or another excavator can be used to compact soils placed back into the trench. Soils removed from the trench that are high in clay are piled in a row along the inside edge of the wetland, between the topsoil and the trench. Soils that contain gravel and sands are placed in a row along the outside edge of the wetland. The trench must begin and end high enough on the slope so that no gravel is found in the bottom, or that the elevation of bedrock in the bottom is equal to the planned elevation of water in the new wetland.

The dozer works to fill and compact suitable soils in the trench while the excavator continues to dig. It fills the trench by pushing soils over the edge, and by pushing soils into the trench down a ramp it forms over the edge. If possible, the trench is dug wide enough so that the dozer can move from side to side, placing a track down the center for compaction. It is most important to compact those soils in the bottom of the trench used to replace the gravels.

In some cases, on level ground and in other locations, it is not possible to begin or end the groundwater dam on higher ground where the subsurface permeable layers run out, on these sites, it may be necessary to build the groundwater dam in a circle around the wetland being built.

If a soft layer of bedrock is present in the bottom of the trench, that layer should be removed. On some sites, between 12 to 18 inches of soft shale are taken from the top of the bedrock layer to expose the much harder impermeable layer below. It is evident that one has dug deep enough when waters flowing across the trench fall vertically into the bottom of the excavation, and then rise in elevation.

Not taking time to build a groundwater dam is like driving well over the speed limit; you might get lucky and arrive faster, but sooner or later, you are going to get caught!

While digging the groundwater dam, collapsing sidewalls can cause problems by allowing permeable soils to span across the trench being built. To reduce cave-ins, the ditch is dug wider at the top and narrower at the bottom. This is known as being "stepped down." The higher the sand and water content in the soil, the wider the top of the ditch must be dug to prevent collapsing. In most situations, the ditch is dug 20 feet wide at the top and 12 feet wide at the bottom.

Avoid piling soils along the edge of the ditch being built. Keeping the base of soil piled away from the edge will reduce cave-ins, and the possibility of gravels oozing back into the trench. Using a large excavator with a longer reach makes it easier to pile soils away from the edge. When using a smaller excavator, it may be necessary to build a low clay dam along the edge of the ditch to keep gravels and sand from falling back into the trench. The dozer may also have to be used to push soils back from the edge as they are removed by the excavator.

When digging a curved trench, it is more efficient to connect a series of short, straight segments rather than attempting to create a curve that has smooth sides. Each time the excavator has to move, it backs up, moves over, and digs another straight segment that connects with the previous segment.

It is critical that sands and gravels do not remain in the trench or that they are allowed to fall into the trench while it is filled. Leaving them in the trench would have the same effect as installing a drainpipe beneath the dam.

Soils are initially pushed into the trench in piles, and then smeared in layers from 12 to 18- inches thick. They are compacted by having one track of the dozer or excavator travel over the same piece of ground at least four times before adding another layer. The bucket of the excavator can also be used to compact soils. It is filled with soil to increase its weight, raised up, and then slammed down to obtain compaction.

Zones of gravel and sand in the trench being dug often appear different in color and texture than the fine-textured impermeable soils used to build the groundwater dam. Sands are often red or orange in color, and gravels are gray or black. The silt-clay used to fill the trench is generally brown. Equipment operators are quick to point out that gravel and sand are soft and easy for the excavator to remove, while clays stick together and take considerably more horsepower to remove from the trench.

Subsurface layers of sand and gravel can vary in thickness, rising and falling, appearing like waves in the ground. These permeable layers can also disappear in sections as the dam is being built. Avoid the temptation not to trench in these sections, as gravel layers can disappear and then reappear over distance.

A dozer must work closely with the excavator digging the trench to fill in sections as the bedrock is exposed. Cleaned sections of trench must not be left open any longer than necessary, as the sides are likely to collapse and place gravel across the bottom. The trench must be filled and compacted with suitable soils to an elevation that is above the sand and gravel layer as it is dug. In this way, should there be a cave in, only suitable soils found above the gravel layer will fall into the trench, and these can be spread and compacted in place.

Where permeable materials are thick and saturated with water, heavy equipment operators must work together to fill short sections of trench as the bedrock is exposed. This may require the use of hand signals and blowing of horns to alert each other when it is time to fill the trench with soil. The excavator operator works to clean a short section of trench, and as the last bucket of gravel is removed, suitable soils are added to replace it, avoiding time for a cave-in of permeable soils.

Standing pools of water in the trench can be removed by scooping them out with the excavator. Where flowing water is present, it must be blocked quickly by placing a large amount of soil in the trench, and then pushing it down with the excavator bucket, or by running over it with the tracks of the dozer to obtain some compaction. When encountering water gushing into a section, it may not be possible to compact the soil in layers as the trench is rapidly filling with water. In these places, large piles of soil are dumped in and the soil is compacted in layers up to 48 inches thick. Remember that the task with the highest priority in these wet places is replacing the permeable soils in the bottom of the trench with suitable soils, not compaction. Working through sections where soils are soft and prone to collapse can be unnerving, and requires close cooperation of everyone involved for success. For safety, it is critical to keep people on foot out of the trench, and the bodies of personnel operating heavy equipment above ground level.

Where great depths are encountered, use the excavator to place and pack soils back in the trench before moving back to dig another section. The larger excavator can use its bucket to place and pack soils in the trench to an elevation which is the same as the top edge of the permeable layer. After this is completed, cave-ins will not compromise the integrity of the groundwater dam, as only soils high in clay can fall into the trench.

There are advantages in using two excavators to build the groundwater dam. The larger excavator digs open the trench, and the smaller machine works in the ditch in front of the larger one to fill with suitable soils. While the initial filling is done by the dozer, the second excavator operator down in the trench has improved visibility for spotting gravels left in the trench, and can rapidly fill the trench by scraping and placing soils in the bottom.

It is best to begin the trench at one end and continue to the other end, as it can be difficult to find the beginning or end of completed sections if they are not finished consecutively. The dam should be constructed over the groundwater dam as it is being built to avoid leaving unfinished sections open for very long. Partially completed sections are low places that collect rain and floodwater. It costs time and money to remove the waters that collect in these low places so they can be finished.

The large excavator is essential when building long and deep groundwater dams. Its huge bucket allows for the efficient removal of large amounts of soil needed for these projects, and its long arm allows for digging deep trenches in a single pass. More importantly, the long reach enables placing soils and gravels removed from the trench away from the edge so they do not fall back in during construction, avoiding the need for the dozer to push these soils away from the edge when working with small equipment. The large capacity bucket is crucial for keeping up with cave-ins while digging through unstable soils. While the cost of hiring a large, 300-Series Excavator may be twice as much per hour as the smaller 100-Series machine, it will move four times the amount of soil in a day.

When building a groundwater dam, it is best that the thickness of gravel layer not exceed the thickness of the overtopping layer of suitable soils on the construction site, as clay soils may have to be trucked in from off site to fill the trench if there is not enough soil on hand to replace gravels that are removed. For budgeting and planning purposes, it is wise to determine the thickness of soil layers beneath the dam's location prior to advertising the contract. This can be done by using an excavator to dig test holes along the length of proposed dam. Providing a sketch to prospective bidders showing the thickness of soil layers and depth to an impermeable layer that will form the base of the groundwater dam is essential to keeping prices low. This data will help contractors develop realistic bids so they do not have to inflate estimates to cover possible conditions they cannot see beneath the surface.

Build the groundwater dam when the water table is lowest during the year. Attempting to build when the water table is high is problematic; waters will soon fill the excavated trench, making it almost impossible to compact soils for the groundwater dam.

On higher elevations in forested areas that were not farmed, it is often possible to use only a dozer to build the wetland. The operator starts digging the trench for the groundwater dam by backing up along the outer edge of the cleared area and then pushing suitable soil inward toward the center of the wetland being built. This action moves soils that are high in clay to a location where they can be efficiently placed back in the trench to build the groundwater dam. Up to a three-foot thick layer of soil is moved this way into the center of the wetland. The operator then changes direction and begins digging a deep trench at one end of the dam which covers the entire location of the dam. The channel is dug to a depth below that of any tree roots, and under the bottom of any crayfish burrows. It is critical to watch the bottom of the trench while the operator is pushing and shifting soil within the channel to make sure that it is dug deep enough. Seeing that tree roots or crayfish burrows are still present in the bottom of the trench indicates it must be dug deeper.

When restoring a wetland on hydric soils, it may be tempting not to construct a groundwater dam. Unfortunately, such a decision may result in failure of the wetland to hold water as planned. While buried drain lines can be present in saturated fields, they are often plugged with soil. These drain lines can be opened by flooding over the top of the field with water that is deeper than normal. An increase in water pressure is all that is needed to cause waters to leave the poorly-constructed wetland by traveling beneath the dam through drain lines flushed clean with flowing water.

Topsoil is piled inside of the planned wetland in a long row for later spreading in the completed wetland.

Topsoil has been stripped from a section of groundwater dam to be built in one day. The truck is parked on a road along the banks of the Dix River in Kentucky.

A large excavator, 300-Series, is used to dig a deep trench down to bedrock. Soils high in clay are piled inside the planned wetland on ground where topsoil was removed.

Groundwater dam construction progression: Large field near river containing deep gravel deposits

A groundwater dam was built around this 70-acre field with a 0.5 percent slope along the Dix River in Lincoln County, Kentucky.

Flagging shows the dozer operator the area of topsoil to remove in advance of the excavator. The distance stripped is equal to the portion of the groundwater dam that can be built in one day.

Gravels and sand are placed outside the planned wetland where topsoil was not removed. The long reach of the excavator allows gravels to be piled far enough away so they will not slide back into the trench.

This shows examples of gravels removed by the excavator.

Top layer of soils which are high in clay are being removed and saved for replacing in the trench.

Bottom layers of soil high in gravel and sand are being removed and placed outside the planned wetland. Bedrock forms the base of the trench. Notice how water pools on top of the bedrock.

Soils in the upper portion of the trench are high in clay and are not collapsing. Notice the buried clay drain line cut by the excavator in the center of the photo.

The excavator cuts into the bedrock, removing the soft upper layer. Little gravel is present in this section. Gravels will occur in waves, with thick and thin sections.

The dozer pushes clay soils into the trench.

The dozer keeps up with the excavator, pushing clay soils into the trench before the walls collapse.

Clay soils are spread in six-inch layers, with each layer being compacted by the dozer by traveling over it three to four times.

A backhoe is used to compact soils in the groundwater dam where bedrock is near the surface.

Clean, washed gravel acts like a drain and care must be taken to completely remove them from the trench.

Removing clean gravel is difficult, as the sides of the trench will cave in rapidly, spanning over and across bedrock. These sections are difficult to fill and compact with a dozer.

In clean gravels, the excavator must work quickly to dig a short section of trench.

The excavator places clay soils in the cleaned trench at the far end of its reach before the sides collapse.

The clay soils are spread across the bottom of the trench and compacted using the bucket on the excavator. This also must be done quickly where waters are entering the trench.

Clay soils are added and compacted with the excavator until they are equal to the elevation of original clay soils on either side of the trench. The dozer is then used to complete the filling and compacting.

Clay soils are placed in the trench on top of bedrock to a level above the elevation of gravels on either side of the trench.

Three pieces of heavy equipment are needed to build groundwater dams that are deep with thick gravel layers.

The large excavator cleans the trench down to bedrock while the second smaller excavator fills the bottom of the trench with clay soils above the elevation of gravel on either side

Clay soils are added quickly to replace wet gravels as they are removed.

The smaller excavator spreads and packs clay soils in the bottom of the trench.

Clay soil is kept in the bucket for weight to increase compaction.

To ensure that gravels were not covered during the filling process, the larger excavator cuts into soils compacted by the smaller excavator for a visual inspection. When the operator sees that no gravels are present in the trench, the next section is filled with clay.

The dozer fills and packs the upper portions of the trench to complete the groundwater dam.

The elevation of the groundwater dam is checked using a laser level at frequent intervals.

The groundwater dam is finished at the base of a hill. It began at the base of the hill on the other side at the same elevation, and continued in an arc around the lower edge of the planned wetland.

Here a groundwater dam is being tied in to the base of the hill where the gravel ends. The trench is deepest at this point.

The dozer backs up to the edge where topsoil was removed and pushes clay soils into the trench to complete the dam.

After the above-ground dam is completed, the gravels, sands, and topsoil are spread by the dozer. The area is purposely left rough and not graded to avoid compaction, which is critical to tree survival and growth.

Both the dozer and the excavator are used to spread gravel, which can be placed in front or behind the dam to make gradual slopes.

On the newly-completed dam and groundwater dam, the slopes are gradual, less than five percent, so the dam will blend with its surroundings. Ridges and mounds are left to increase plant diversity in and around the new wetland.

Within the large field surrounded by a dam and groundwater dam, the dozer shapes a depression for a smaller and deeper wetland. Grading is being done under freezing conditions to avoid compaction, which will be critical to future plant growth.

The dozer and the excavator place a snag following completion of the dam and the groundwater dam. Both pieces of equipment are used to prevent the tree from falling towards an operator.

The excavator breaks large woody debris to size before placing it in the restored wetland. Broken logs appear natural and provide better habitat compared to cutting with a chainsaw. (Cynthia Sandeno photo)

The excavator operator uses the bucket to anchor snag in a wetland restored in a floodplain. Notice how rough the soils are left around the wetland. (Cynthia Sandeno photo)

Building the Dam

The dam for the wetland is built over the top of the groundwater dam, using soils generally obtained from the area that will be flooded by the new wetland. The dozer scrapes and pushes up layers of soil from the zone between where the topsoil was piled and where the dam will be built.

The dam should be built out of mineral soil that has a mixture of silt and clay which will form a thin ribbon at least two inches long after adding water as illustrated earlier. The dam is constructed in layers approximately 12 inches thick, where each layer is compacted by running over it four or more times with one track of the dozer before the next layer is added. Roots, branches, topsoil, sand, and gravel are to be kept out of the dam. It is especially important to build the center of the dam from soils that are high in clay so it will not leak.

Avoid creating a deep trench in front of the dam by instructing the dozer operator to back to the uphill edge of the cleared area to begin scraping soil for the dam. Deep water along the base of the dam can cause problems from muskrat burrowing, will decrease plant diversity, and will make the wetland appear artificial. For longer dams on larger wetlands, the dozer should back up to the topsoil pile, which should have been placed 200 feet from the dam, to begin moving soil for the dam.

Begin by pushing up plenty of soil for the dam. After several layers have been compacted, use wire florescent flags or stakes to mark the planned elevation for the top of the dam. To help the dozer operator know the height of the dam to build, the top of the flags should be set equal to the planned elevation for the top of the dam. The dozer operator will later grade the top of the dam, and push any excess soil onto the front or back slope. It is permissible to have too much soil piled for the dam, as this can help avoid low spots, and will provide the soil needed to place gradual slopes on the front and back of the dam.

Soil should be removed leaving an irregular surface to give the wetland a natural appearance. When installing a

A dozer spreads excess soil to place a gradual slope on the back of the dam.

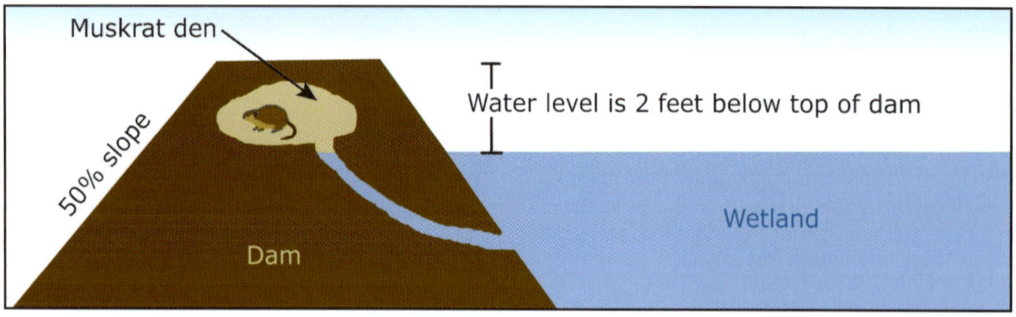

Muskrats are likely to damage this dam

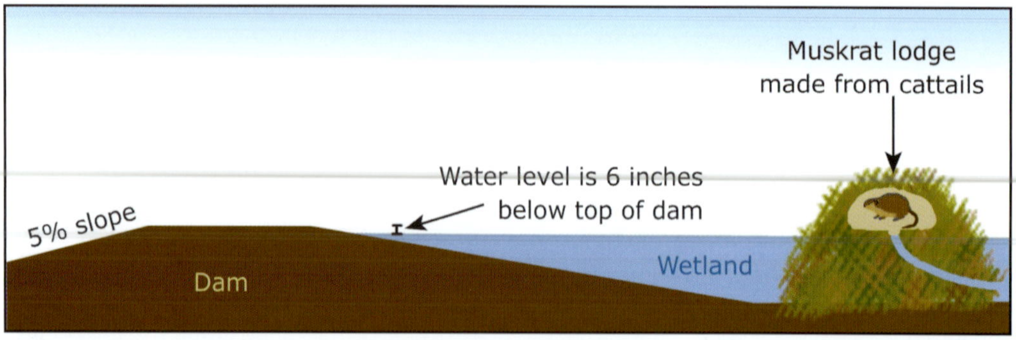

This dam is resistant to muskrat damage

drainpipe, the bottom of the wetland should be shaped so that waters flow to the inlet when the wetland is drained. Place gradual slopes along the edges of the wetland to create a variety of water depths. Gentle slopes provide conditions for a variety of plants, and reduce the possibility of muskrats tunneling into the dam. It is not necessary or good biological practice to level the bottom of the wetland when the dam is completed. Leaving dips and bumps will help increase plant and animal diversity.

Almost any texture of soil from clay, silt, to gravel, including topsoil, may be used to build the inside slope or back slope of the dam. Gravels removed during construction of the groundwater dam may be disposed of by creating a gentle back slope on the dam, or they can be used to place a layer over the top and inside slope of the dam to discourage muskrats from tunneling into the dam, as shown in these drawings.

The more gradual the slopes placed on the dam, the more natural the wetland will appear, and the less maintenance it will require. Slopes should be kept to five or less percent to eliminate the need for mowing and to reduce muskrat damage. Once the dam and spillway have been completed, any topsoil that was saved can be spread over the bottom of the wetland and up over the dam.

When building a dam that will take longer than one day to complete, it is wise to be prepared for rain. A thunderstorm can saturate soils and shut down construction for a week. Make sure the ditches in and around the worksite are kept open during construction, and that new ditches are dug to prevent waters from collecting on the site if it rains. Watch the weather forecast and direct the contractor to reopen blocked ditches prior to rain. While an effort to maintain ditches takes only minutes of work, it can save days of delay.

During the final stages of dam construction, steps can be taken to make it appear more natural. These can include adding soils to the front and back of the dam to form peninsulas and waves, and creating islands to soften the otherwise straight line and artificial appearance of the dam.

The dam should be seeded and mulched with straw to prevent erosion and to reduce colonization by invasive species. Seeding exposed soils with native species will increase plant diversity in and around the new wetland. While seeding with hardy species like fescue and perennial rye may seem like a good idea, using them can prevent native wildflowers, trees, and shrubs from growing for years. Because native grasses and wildflowers can take longer to germinate, winter wheat or oats should also be sown to prevent erosion. Wheat will germinate in only three days after rain, and is non-invasive. When sown at a rate of 100 pounds per acre or less, wheat or oats will act as a nurse crop for native plants.

Volunteers spread native willow oak leaves over exposed soils to prevent erosion in a newly-constructed wetland at Sweet Briar College in Virginia. (Mike Hayslett photo)

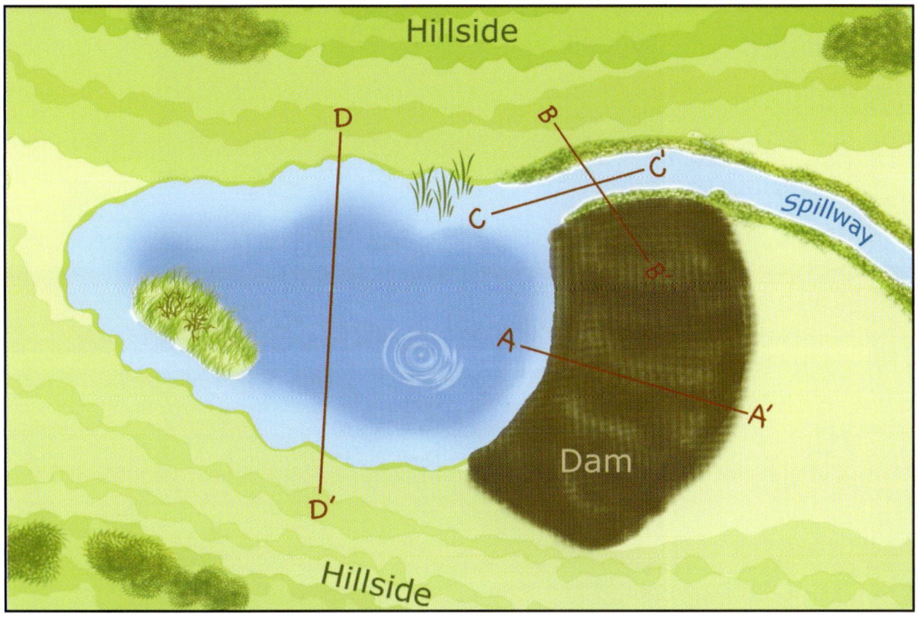

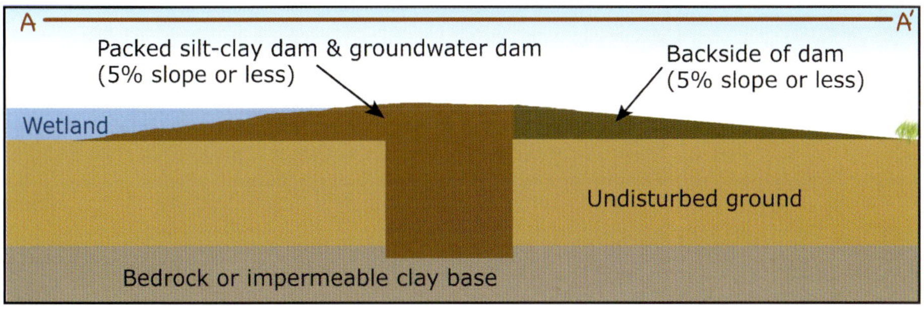

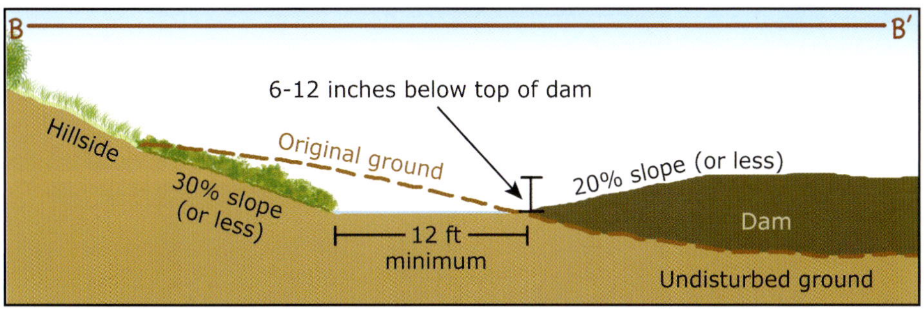

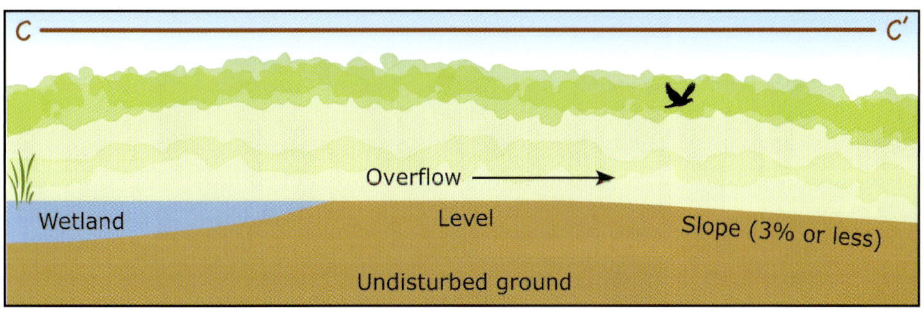

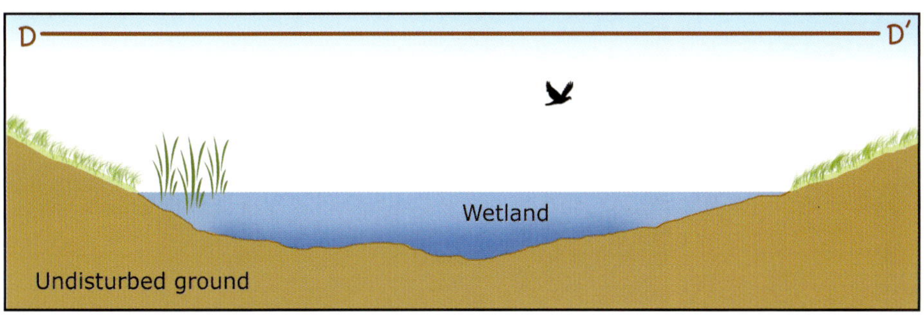

Wetland construction, Surface Water Technique, Plan view (Left)

Groundwater dam and dam construction, Profile view (Left A)

Spillway construction, Profile view (Left B)

Spillway construction, Profile view (Left C)

Wetland pool, Profile view (Left D)

Building on Steep Ground

Wetlands constructed on steep slopes often look unnatural because of high dams with steep slopes. Steps can be taken to build naturally-appearing wetlands on hillsides, and on level ground along the edge of a steep slope.

A wetland will appear more natural on a steep slope if more excavation is done to create a shallow depression rather than mounding up soils to back water uphill. Avoid building a high dam, as soils piled to create the dam will make it appear more visible, especially when the backside of the dam is much steeper than the original slope on which the wetland is being built.

Hide soils removed to create the depression for the wetland by pushing them downhill in a layer over the existing slope. The goal should be to match the slope created for the backside of the dam with the slope on which the wetland was built.

This emergent wetland was created on a steep slope along Muse Branch on the Daniel Boone National Forest in Kentucky. The dam on the left was built far enough away from the trees to allow room for gradual slopes.

Topsoil is placed in the new wetland. This is a key component that will jump-start the wetland's vegetative growth.

Roots were left on trees to provide improved habitat for plants and animals.

To reduce disturbance beyond flags the dozer back-blades piles of topsoil and vegetation into the new wetland.

Exposed soils are seeded to wheat and mulched with wheat straw to prevent erosion, and reduce potential colonization by non-native, invasive plants. An area 80 feet in diameter was disturbed to build a wetland that will be approximately 40 feet in diameter.

A spillway is created to provide a gradual path for waters to flow from the wetland. Notice how the dam with its gradual slopes occupies two-thirds of the cleared area.

Root wads are placed at random in the wetland, along with small mounds of topsoil and vegetation scattered over compacted clay soils.

Large trees are placed in and around the wetland to prevent motorized vehicles from driving on the dam or in the wetland.

Spillway Construction

Natural wetlands sustained by runoff generally have a low place or spillway along their perimeter whereby water can exit under flood conditions. Because both restored and created wetlands will fill with water at some time, it is important to direct how waters leave the wetland to prevent damage and erosion. A spillway can be a natural low place or a constructed stream that provides a path for waters to flow out of the wetland without causing damage to the dam. The spillway is a safeguard, like the overflow drain on a bathtub. Imagine the spillway as a road that has a hardened surface that carries excess waters around the dam.

The entrance of the spillway is set at an elevation lower than the top of the dam, and follows a gentle path downhill across undisturbed soil. The spillway is generally constructed 12 or more feet wide, and follows a downhill slope of three or less percent. To prevent erosion, water flowing down the spillway should be directed over soils not disturbed by construction, or the surface should be hardened with rock.

In areas where beaver are present, and on wetlands with a large watershed, it is wise to construct a much wider spillway, over 24 feet across, in order to reduce the depth

The spillway on this new wetland guides runoff down a gradual 2 percent path. The wetland was built on a steep 6 percent slope. The spillway is located between the dam and the base of the hill.

and velocity of water flowing from the wetland. A wider spillway will diffuse water flow into a thin sheet, which beaver are not as likely to block in comparison to a narrow channel. A spillway should be placed on all wetlands that are built on a slope, regardless of size, to prevent loss of the dam. A bonus to creating a wide, gradually sloping spillway is that the waters flowing down its winding course will eventually develop into a naturally-appearing stream with wetland characteristics.

A spillway is usually located at one end of the dam, where it ties into higher ground. The spillway is created after the dam is built, to ensure its entrance is lower than the top of the finished dam. Because a functioning spillway is often wet, its optimum placement is opposite the primary access to the dam, so that the wetland can be visited and enjoyed without day trippers having to wear waders. Use a level and rod to follow the potential path that water will take after leaving the wetland to make sure that waters flowing from the spillway do not affect neighboring property or other improvements. It may be necessary to create a long path for water to follow to avoid causing problems.

The entrance of the spillway should be set from 6 to 12 inches below the top of the dam. Since few earthen dams are built perfectly level, it is always a good idea to use a survey rod and level to check elevations along the top of the dam before creating the spillway. Do this by walking along the top of the dam and stopping every 20 feet or less to record elevation. Any place along the top of the dam that appears to be low should be checked. Keep track of the lowest elevation found along the top of the dam, and then mark the entrance of the spillway to be from 6 to 12 inches below this elevation. This is an example of how to determine the correct elevation to set the spillway. For a 400-foot long dam, the following elevations are recorded along the top of the dam: 6.0, 6.1, 6.0, 6.1, 5.9, 6.1, 6.1, and 6.0, and the elevation originally set for the top of the dam reads 6.0. The elevation of the inlet for the spillway should be set at 7.1 (6.1 + 1.0). This figure is reached by taking the lowest measurement found on the top of the dam-- 6.1-- and adding 12 inches. Water should leave the wetland when it is 12 inches below this height.

It is important to be on-site when constructing the spillway to ensure that it is in the best location, has the correct elevation, and a gradual slope, as wetland failure can result from improper construction of the spillway. The inlet of the spillway can be viewed as a separate, lower dam that water should pass over before it will breach the constructed dam. Determining the elevation of the entrance of the spillway is critical to protecting the dam. When set too low, much of the dam will be above water, giving muskrats room to tunnel and cause damage. When set too high, the dam can become saturated and too soft for mowing, or waters may flow over and cause erosion. Calculating the height and building the spillway takes time, and should not be rushed.

Start building the spillway after the top of the dam has been graded by placing a wire florescent flag in the ground to mark the entrance of the spillway. The flag should be set in the ground at the peak elevation for the spillway so that the equipment operator can use it as a guide. Ask the equipment operator to scrape a downhill path, beginning where the flag touches the ground. The path should start level, and then continue downhill on a gradual slope of less than three percent. Ask the operator to begin by only removing a little soil at first, and then stop him or her frequently to check the elevation at the entrance of the spillway. Also check the path that water will follow to confirm that it goes downhill gradually. It is best to have the equipment operator lower the soil an inch or so at a time rather than gouge it out in one pass. Removing a little soil helps keep the elevation of the spillway from becoming too low, and from sloping too steeply. Removing too much soil means that the spillway will have to be backfilled. This takes more time, and will result in poorer performance, as the soil placed back into the spillway is soft and more likely to erode. Do everything possible to avoid removing more soil than is necessary; the bed of the spillway should be native, undisturbed soil. Set up the level close by and use it often during the operation. Any surplus soil removed to create the spillway can be spread along the backside of the dam.

To save the expense of having to protect a spillway with rock, build wetlands in locations that have small watersheds, and avoid intercepting waters from a perennial stream unless you have experience in working with flowing water. Even with these precautions, it is sometimes necessary to armor the spillway to prevent erosion. Rock may be needed when waters flow down a spillway over steep ground or across soil not protected by vegetation. Rock is also needed when the spillway is expected to carry water on a regular basis. The spillway must be dug deeply enough to allow room for the rock. The rock has to be imbedded into the ground so that moving waters do not roll it downhill and cause erosion.

Constructing a rock-lined spillway is much the same as creating a rock-lined ditch, except the surface of the rock needs to be set at the same elevation as the spillway. Rocks of various sizes ranging from two to eight inches across should be packed into a surface resembling a hard road for the spillway. It works well to use a mixture of rock sizes, as they will pack and remain in place better than rocks of a uniform size. The rock needed for a spillway construction is generally ordered from a quarry and is delivered by dump truck. Contractors often call the size of rock needed for a spillway "channel liner". It is also a good idea to mix some soil in with the rock so that plants will grow in the spillway. These plants will eventually mask the rock.

For example, one wants to create a rocked spillway whose final elevation is 12 inches below the top of the dam. The elevation of the entrance of the spillway is excavated to be 12 inches below the top of the dam. Add an additional amount equal to the thickness of rock placed over the spillway, which, in this case, is estimated to be eight inches. When completed, the final elevation at the entrance of the spillway lined with rock will be 12 inches below the top of the dam.

The expense of lining a spillway with rock can be saved by designing a wide path that carries waters over undisturbed soil down a gradual slope.

To construct a 12-foot wide spillway that will be protected by an 8-inch thick layer of rock, plan on ordering approximately 30 cubic yards of rock for every 100-foot length of spillway to be built. When in doubt about whether or not rock is needed to protect the spillway, it is often better to use it at the time of construction. Once the wetland has filled and waters have begun flowing down the spillway, it is more difficult and costly to repair the damaged spillway.

Using both a dozer and an excavator is an excellent way to build a spillway. The dozer creates a path for the spillway, and the excavator places rocks in the spillway. The dozer

Massive erosion is occurring in this spillway constructed on a steep slope with livestock access.

This wide dam constructed with gradual slopes became a wet-meadow wetland.

Sedges and trees grow naturally on this wide wetland dam, which was hand-sown to wheat and mulched with wheat straw following construction.

or excavator can then be used to pack the rocks into the spillway so they will stay in place.

It is possible to build a wetland without a constructed spillway, but the dam must be wide and gradual, with slopes less than one percent to prevent washing. The main advantage in allowing waters to seep over and into the dam is that the dam will also become wetland. Waters leaving the wetland will flow over the dam in a sheet-like pattern, saturating the surface and creating conditions suitable for development of wet-meadow, shrub, and forested wetland. For best results, a layer of topsoil mixed with vegetation, and not compacted should be spread over the completed dam. One consideration when deciding whether to build a spillway is that these wetland dams, as a rule, are too soft to support rubber-tired vehicles without getting stuck, and causing damage to the wetland.

Handling Topsoil

Wetland construction projects can generate large piles of topsoil. Deciding what to do with it can affect the success and cost of a wetland project. Spreading a loose layer of topsoil over the finished wetland will improve conditions for the growth of plants compared to leaving compacted mineral soils in the bottom. Topsoil can also contain the seeds of wetland plants that have lain dormant for years, and with inundation, these seeds may germinate and grow in the restored wetland.

Adding a thick layer of topsoil to the bottom of a wetland can provide amphibians and reptiles with a safe place to spend the winter.[37] Turtles will bury into the bottom of a wetland for hibernation and under drought conditions, so adding from 12 to 24 inches of loose soil to the bottom of a wetland before it is filled can help them evade detection from predators such as the otter.

This young snapping turtle was found buried in the soft mud of a restored wet-meadow wetland on the author's farm in Rowan County, Kentucky on July 4, 2008.

Spreading topsoil over mineral soil exposed above the water level also has advantages. Seeds planted in the topsoil will grow better than those sown on mineral soil, and the topsoil probably will contain many seeds of its own that will germinate with rain. However, there can be disadvantages should the topsoil contain invasive plants, such as Johnson grass or Japanese knotweed that will be replanted if the infected soil is spread.

While it is desirable to return all the topsoil removed during construction into the wetland, construction costs may be lowered by spreading a layer about four inches thick in the bottom of the wetland, and then using the remaining topsoil to form gradual slopes on the dam. Excess topsoil can also be shaped into islands for planting trees.

One does not have to add topsoil to create a successful wetland. By their nature, wetlands serve as a collection basin for soils and organic material from surrounding lands. For example, it is not uncommon to find over 12 inches of organic material accumulated in the bottom of deep farm ponds created 30 years ago when topsoil was considered waste following construction. Wetlands built from gravel pits or borrow areas devoid of topsoil will contain a thin layer of topsoil ten years after creation, and can show an impressive variety of aquatic plants.

In some cases, the surface of the original wetland can lay buried deep beneath the top on the restoration site. This may occur along major rivers where significant quantities of soil have been deposited since European settlement. To determine if a wetland has been buried, holes should be dug to see if a historic layer of topsoil lays buried at the site. Sometimes a soil auger will work, but often an excavator is needed to make this verification. Should a buried topsoil layer be found, a decision should be made if the budget for the project is great enough to remove the upper layers to expose the wetland to its original elevation. In the majority of situations, it is not necessary to remove deposits to construct a successful wetland. A choice often can be made whether it is better to create a number of naturally-appearing wetlands on top of the filled surface or to remove the sediments to restore one wetland at the historic elevation.

Recommended Equipment For Building Groundwater Dams

Proposed Wetland Specifics:
Length of dam is 1,000 feet or less
Groundwater dam will be 12 feet deep or less
Absent or thin buried gravel layer
Estimated cost per foot = $10.00

Excavator: 100 series (or larger); metal tracks; 42-inch wide bucket; 80 HP minimum; 22,500 lbs minimum; 10 years old or less

Dozer: 105 HP minimum; 20,000 lbs minimum; 10 years old or less

Proposed Wetland Specifics:
Length of dam is 1,000 feet or more
Groundwater dam will be 12 feet deep or more
Thick buried gravel layers present
Estimated cost per foot = $20.00

Excavator: 300 series; metal tracks; 60-inch wide bucket; 200 HP minimum; 10 years old or less

Dozer: 130 HP minimum; 28,000 lbs minimum; 10 years old or less

Avoiding Flood Damage

Flooding can damage wetlands unless steps are taken during construction to protect them from flowing water. Signs of flooding can include grasses caught in fences and hanging from branches, logs and trash scattered in fields, and plants wrapped around the bases of tree trunks.

It may not be necessary to place a large overflow pipe or cement spillway on a wetland completed in a floodplain in order to prevent damage from floodwaters. These measures are often unnecessary because under flood conditions the entire wetland, including the dam, will be inundated. What is most important is to shape the wetland and dam so that it can survive being overtopped by moving water.

The primary method of protecting a wetland from flood damage is to place gradual slopes on both sides of the dam, and to avoid sharp turns and vertical edges that cause turbulence. Building slopes five or more percent gradual on the front and back of the dam will allow waters to flow over them without causing damage.

This two-year-old emergent wetland was built from a log landing on a mountain ridge on the Daniel Boone National Forest by shaping and compacting soils high in clay.

It is critical to obtain a rapid and dense growth of vegetation on the dam to protect it from washing when building in a floodplain. Exposed soils on sections of the dam should be seeded to wheat and mulched with straw as they are completed. One heavy rain will wash fine soils off the dam and make it exceedingly difficult to establish plants unless soils are protected by a covering of straw. Extensive damage can be caused to newly-constructed wetlands from heavy rains wrought by hurricanes and thunderstorms, so it is important to help plants take root so they can hold soils.

To avoid damage and ensure longevity when building in a floodplain, the profile of the wetland should be lowered to include more excavation and less dam building. High dams in a floodplain require frequent maintenance and are prone to failure. Wetlands with low dams and gradual slopes containing pools below the surrounding floodplain should last for years.

Obtaining Suitable Compaction

Wetlands designed to hold rainfall and runoff generally have soils in the dam and groundwater dam that are high in clay and compacted. Natural soils not disturbed by heavy equipment or livestock contain air spaces whereby rainfall can soak into the ground. These soils are considered well structured and transmit water considerable distances until it reaches the water table or enters a stream or lake. Even fine-textured soils that contain high amounts of clay are permeable and can transport large quantities of water.[38] Compaction causes a reorientation of soil particles that reduces the volume of air in the soil, which slows water movement into the ground.[39] Compacted soils form a barrier to water movement, and are said to be impermeable.

Once compacted, soils can remain impervious to water for hundreds, if not thousands, of years. Compaction is the reason all the soil removed to dig a hole for a fence post can be placed back in the hole around the post when tamped with an iron bar. Compaction explains why ruts created by rubber-tired vehicles will hold water like mini-wetlands. A lack of compaction explains why hand-dug holes rarely contain water for more than a day following a heavy rain.

Constructed surface water wetlands can fail to hold water if soils are not compacted during construction. There are a number of ways to determine if soils are being compacted when constructing a wetland:

- A hard ball can be formed by hand that will not break apart when dropped.
- A thin ribbon or tube of earth can be formed by hand without adding water.
- Equipment tracks and vehicle tires will stay on top of the soil and not sink after repeated passes.
- It is difficult to dig a hole by hand with a shovel.
- Soils have a smooth, hard surface when cut with a dozer blade, not tearing or appearing grainy.

Suitable compaction can generally be obtained by requiring the tracked dozer or excavator to travel over each layer of soil four times or more as it is spread to form the dam or groundwater dam. Soils should be spread in layers of no more than 6 inches thick before being compacted. Equipment with rubber tires will provide greater compaction than tracked machines. When available, use a backhoe with its rubber tires to travel over each layer of soil placed by tracked equipment to obtain greater compaction.

Civil engineers often use a special gauge called a penatrometer to measure compaction when building roads and parking lots. They aim for 90 to 95 percent compaction, which works well for wetland construction. However, when building a wetland with moist soils, it is often not possible, nor is it necessary, to obtain this high degree of compaction. In many situations, one will have to do one's best to shape the basin and dam with a dozer and excavator, and then spread and back drag the soils to form the slopes of the dam. Even though these dams built from wet soils can be expected to settle over time, they should hold water as planned. When building a wetland from wet soils, plan on returning to the site after the dam has dried with heavy equipment to fill and level sections where settling has occurred.

Water from a fire pumper truck was added to the loose soils. (Richard Hunter photo)

The greatest problem in obtaining compaction occurs when building wetlands from soils that are too dry. Wetlands built from dry soils that are high in silt fail when waters soak through the dam and groundwater dam. Dams constructed of dry soils can settle and crack, causing washouts. There are a number of ways to tell when soils are too dry for building wetlands. The first is when soils removed from a test hole on site cannot be shaped into a thin ribbon or tube. The second is when dust is observed blowing across the work area when equipment disturbs the ground. A third method is when lumps of soil removed from the ground crumble instead of clumping when crushed by heavy equipment.

Puddles show how much water was needed to moisten the dry soils. The moist soils were then mixed and compacted by using the dozer. (Richard Hunter photo)

Surface water wetlands constructed from soils high in silt should be completed when ground conditions are moist in order to ensure compaction. In some locales, spring and winter are the best times of year to build from silt-loam soils. Silty soils over gravel will drain quickly, and can be worked before clay soils that drain slowly.

A heavy rubber-tired vehicle was used to further compact the moistened soil. Another layer of dry soil was then placed over the top of the compacted soil, and this process was repeated until 24 inches of compacted soils were obtained across the basin. (Richard Hunter photo)

It may be possible to wet soils with water pumped from a nearby stream or pond, or from an available fire hydrant to create conditions suitable for compaction. However, when working in a remote area, it may be best to wait for a rain, returning to mix the wetter top layers with the drier mineral soil to increase its moisture content.

Here a six-inch layer of clay soils too dry to compact were loosened in the planned wetland basin by using the blade on the dozer. (Richard Hunter photo)

An excavator is used to moisten dry soils so they can then be spread in a depression and be compacted in layers.

Woody Debris

Logs placed in wetlands provide sites for turtles and snakes to bask, perches for birds to hunt from and to rest on, and moist protective cover under which the marbled salamander will lay its eggs in the fall when the wetland is dry. There are a number of ways to return large woody debris into a restored wetland. A dozer with a chain can be used to drag logs into the wetland. An excavator can use its bucket to pick up and place logs, or several trees adjacent to the wetland can be cut or pushed into the wetland.

Large logs placed in this restored wetland increase wildlife viewing opportunities by providing perches for birds and basking sites for turtles.

Where vertical snags are desired, consider using an excavator to transplant trees to stand upright in a wetland. Leave the large root mass attached to the tree, and bury it about twice as deep as originally found to ensure stability. Avoid cutting trees with a chainsaw to make the woody debris appear more natural. The broken and splintered ends of logs and branches will provide better habitat than smooth, cut surfaces.

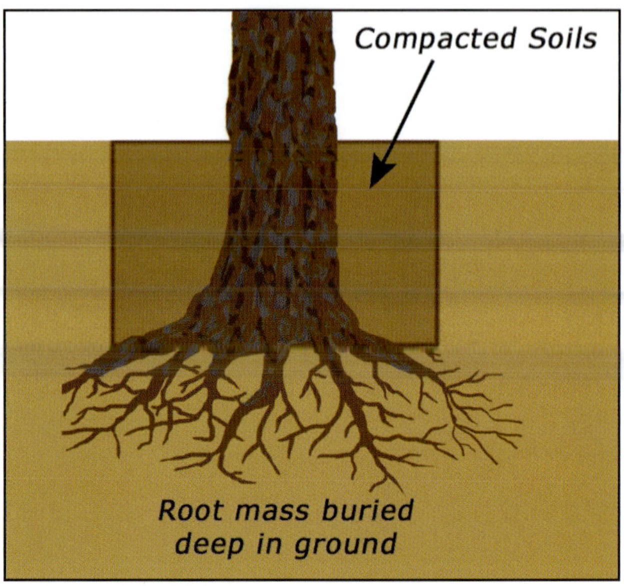

How to anchor dead standing trees

Building in Roads

There are many reasons to consider building wetlands in roads:

- Restores wetlands impacted by road construction
- Improves habitat for endangered species, such as the Indiana bat
- Replaces road ruts being used unsuccessfully by amphibians for breeding
- Helps close roads to unwanted vehicle traffic
- Cleans stormwater runoff
- Controls erosion
- Recharges groundwater

Due to steep slopes and/or the presence of archeological sites, old roads may be the only place available for building wetlands in some locations.

Steps for building a wetland in a road:

- Locate sections of road containing soils high in clay that are 30 feet or longer. Suitable locations often contain mud and ruts, and these ruts may contain water. Use colored flags to mark the beginning and end of the road section to be treated.
- Use a dozer to scrape gravel, vegetation, and topsoil from the surface. Push and save vegetation and topsoil into a pile on the road.
- Expand the area cleared with the dozer to the original width of the constructed road if possible.
- Dig a deep channel down the center of the road that cuts through tree roots and buried drain lines. The bottom of the channel should be based on solid rock or clay soils that do not contain organic material. Push these clay soils downhill and pile in front of the topsoil in the road.
- Turn the dozer around and begin spreading clay soils back into the channel. Place the soils in layers six inches thick, running over each layer at least four times with the dozer for compaction.
- Stop the dozer operator from time to time to check the depth of water in the center of the depression with a construction level. The depression should be deepest in the middle, between the flags, and slope uphill gradually on either end.
- Slope the sides of the channel with the dozer. The slopes can also be made from compacted clay to reduce seepage into surrounding soils.
- Spread any excess soils downhill over the road surface like a dam with gradual slopes.
- Spread the pile of topsoil and organic material over the disturbed area and the bottom of the depression.
- Place large and small woody debris in the depression and on the road.
- Seed with wheat and native species, and then mulch with straw, to prevent erosion.
- Construct a spillway when building on a slope. This is imperative to prevent water from flowing over the dam

and down the road, which will cause erosion.

Wetlands can also be built in roads where springs are present or the water table is high. Clay soils are not necessary, and compaction is avoided when building wetlands in roads to be supplied by groundwater.

This is a small permanent water wetland created from a logging road rut in Morgan County, Kentucky.

This ephemeral wetland was created in a logging access road in Morgan County, Kentucky.

Ruts filled with water show where it is often possible to build a small wetland.

This shows the appearance of the logging road before construction begins.

A dozer removes vegetation and topsoil.

Topsoil is pushed out of the way for later placement.

Mineral soils high in clay are removed to create a deep depression.

The depression must be dug deep enough to cut through all tree roots.

This depression is being dug 40 inches deep in the center.

The piled clay soils are spread back in the depression in thin layers.

The dozer travels over each layer four or more times for compaction.

Topsoil is spread over the compacted depression, now only 16 inches deep.

A spillway is graded for excess waters to leave the wetland, large woody debris is placed, and the wetland is completed in the logging road.

Joe Metzmeier, U.S. Forest Service Wildlife Biologist (shown here), shows how road ruts can be shaped into small wetlands and then filled with woody debris to successfully close roads to vehicle use.

Other Situations & Methods

Clay Liner Techniques

One method of building a wetland that will be supplied with surface runoff, where soils high in clay occur over gravel, is to remove the clay, shape a depression for the wetland, and then return the clay, packing it in layers. This technique allows for the construction of wetlands on sites where a groundwater dam cannot be built, or where a synthetic liner is not available.

The backhoe is compacting soil in the bottom of the wetland as the dozer stands by ready to spread the next layer of soil. Eventually, a compacted layer of clay two feet thick will be created beneath the entire basin. Here, construction is taking place in the winter when soil moisture is high enough to obtain compaction. (Mike Hayslett photo)

In this example, a wetland with a maximum depth of 18 inches is being built on a site with six inches of topsoil over four feet of silt-clay overlying gravel.

Clay on-site

- Mark the area to be cleared with flagging. There should be no more than a three-foot change in elevation from the upper to lower edge.
- Clear trees and shrubs, and temporarily pile near the edge of the work area.
- Remove topsoil, and pile adjacent to work area.
- Remove up to four feet of soils that are high in clay, and place in separate piles adjacent to wetland project. Avoid digging into gravels or placing gravel into the clay piles.
- Remove gravels to create a depression that is three feet deeper than the desired depth of the wetland being built. The entire depression for the wetland must be dug three feet deeper to make room for compacted clays (24 inches), gravel (six inches), and topsoil (six inches) being placed back in the depression. Enough gravel should be saved so that a six-inch thick layer can be placed over the compacted clays to protect them from crayfish burrowing.
- Place soils high in clay into the depression in layers six inches or less thick. Compact each layer by traveling over it at least four times with heavy equipment. Keep adding layers and packing them until at least two feet of compacted clay has been placed over the entire depression.
- Place a layer of gravel at least six inches thick over the compacted clays to protect them from puncture. Compact this gravel layer.
- Spread topsoil over the entire wetland, and do not compact.
- Return woody debris to the wetland.
- Blend remaining gravels into the slope downhill from the wetland to make the dam appear natural, or haul them away.

Because this technique involves moving a large quantity of soil, a dozer at least 105 HP and 20,000 lbs is needed for the job. In addition, open space must be available along the edge for piling large amounts of soil to be removed.

Where clay soils are not present, and a liner cannot be used, it is possible to transport clay to build a small wetland. A thick layer of compacted clay can be used to build a wetland that has a permeable substrate, as long as the clay is protected from puncture. The following steps are used to build a wetland over sand and gravel, using clay where the water table is far below the surface.

Clay off-site:

- Remove vegetation from the planned wetland area.
- Remove and save topsoil.
- Remove permeable soils from planned wetland area to a calculated depth below the surface. Depth of excavation = (Planned maximum depth of water) + (24 inches for compacted clay) + (topsoil thickness) + (six inches for gravel to be placed over compacted clay).
- Place clay in excavation in six-inch or less lifts, packing each layer before more clay is added. A minimum thickness of 24 inches of compacted clay should cover the bottom of the planned wetland.
- Add, shape, and compact additional clay to form the inside slopes of the wetland.
- Spread a layer of rock at least six inches thick over the compacted clay to protect it from puncture.
- Spread topsoil over the rock.
- Place large and small woody debris in wetland, seed, plant, and mulch.

Example:

Planned wetland size = 40 feet in diameter

Planned water depth in wetland = 1.5 feet

Size of excavation needed = 40 x 40 feet

Depth of excavation = 4.5-feet = (1.5 feet desired depth) + (0.5 feet topsoil) + (0.5 feet protective gravel layer) + (2 feet for compacted clay)

Soil to be moved for excavation = 269 cubic yards = (4.5 feet deep x 40 feet wide x 40 feet long = 1.5 yards x 13.3 yards x 13.3 yards)

Clay needed for wetland = 230 cubic yards = (4 feet deep x 40 feet wide x 40 feet long = 1.3 yards x 13.3 yards x 13.3 yards) *Note: this is not an error; 4 feet of loose, hauled clay will compact to a layer 2 feet thick.*

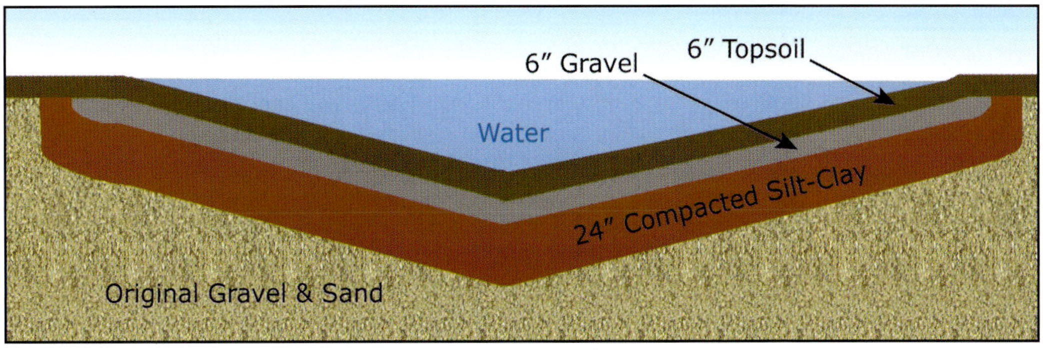

Profile view showing how a wetland can be constructed using a clay liner

A large tri-axle dump truck can haul approximately 13 cubic yards of loose soil, so approximately 18 loads of clay would be needed to build this wetland.

Construction Fill

It is very difficult to construct a wetland from construction fill. The mixture of soil textures, rocks, and debris typically placed in fill are difficult to pack, and therefore, prone to leak. Fill often appears to be a patchwork of different colors when uncovered during construction, not occurring in the uniformly colored layers found in undisturbed soils. Finding buried cans, broken glass, asphalt or concrete chunks provides a clear sign the area was filled. One can generally assume that in construction fill has been placed around buildings, and in the large, level areas near buildings, especially in low or hilly locations.

Building a successful wetland from construction fill requires special attention. The practice has a high risk of failure that can be reduced if constant monitoring occurs during construction. The most reliable technique to use when building over fill is to use a synthetic liner. However, on some locations, larger wetlands can be built at a lower cost when specialized techniques are used to shape soils in place.

In this first example, construction fill consisting of coarse-textured soils, mainly sands and gravels are found over native soils that are high in clay. This is often the case where a surface water wetland was historically buried beneath mixed soils. Basically, the technique involves removing the construction fill, building a deep basin from the buried clay soils, and then partially filling the basin with construction fill. This same technique will work where sandy-loam soils occur naturally over clays.

Building where coarse-textured soil lies over clay:

- Mark the area to be cleared with flagging. There should be no more than a three-foot change in elevation from the upper to lower edge.
- Clear trees and shrubs, and temporarily pile near the edge of the work area.
- Remove topsoil, and pile adjacent to work area.
- Remove construction fill, and place in piles adjacent to wetland being built.
- Build a groundwater dam and dam from native soils that are high in clay. Increase the elevation of the top of the dam to compensate for the thickness of coarse-textured soils and topsoil to be placed back in the wetland. The dam can be built from clay with steep slopes, later placing construction fill over the front and back of the dam to create more gradual slopes.
- Place fill in the wetland, spread excess soils down slope from the dam, or haul it away,
- Spread topsoil over the wetland, but do not compact.
- Replace woody debris in and around the wetland.

This procedure can produce a much larger wetland for the same price of a smaller one built with a synthetic liner. However, much room is needed to pile fill, and construction may be too expensive if course-textured soils are more than two feet thick.

A second example involves a case where soils high in clay were placed in a wetland basin whose soils were high in silt, sand, or gravel. Attention must be given to saving the clay and to obtaining suitable compaction of clays in the groundwater dam and dam during construction. This is best accomplished by monitoring soil moisture conditions, and by taking steps to obtain compaction by using a roller or rubber-tired tractor or backhoe. It is important to be on-site monitoring soil texture and moisture during construction. Layers of rock and coarse-textured soils uncovered must be moved and not placed in the dam or groundwater dam. It is also wise to add additional runoff to make up for possible seepage into the ground.

Building on Wet Ground

Sometimes it is necessary to drain an area before it can be restored. Soft ground and water can make it difficult for heavy equipment to operate, and mud cannot be shaped or packed into a dam. Soils are too wet for construction when the tracks on heavy equipment sink into the ground, and when machines become stuck. Specialized equipment such as wide-track dozers can help when working in wet soils, however, it is generally more efficient to further drain the area so that equipment can be used to shape and compact soils to form the dam.

Drainage of saturated soils that are high in clay allows for the construction of much larger wetlands at a lower price in comparison to digging depressions that will expose a high water table. Once soils are dried, a dam can be built that will inundate many more acres than can be flooded by excavation alone. Creating a groundwater wetland involves removing large quantities of soil to produce a relatively small area of water, while the opposite is true when a dam can be used to build a surface water wetland.

Some managers have trouble accepting that at times wetland restoration sites will require further drainage before restoration can begin. What often happens on agricultural land is that soils become saturated because of blocked drainage ditches and collapsed drain lines. Some say that these wet areas should be left alone because they show wetland characteristics. What these workers fail to realize is the wetland observed is not the same as the one that historically occurred on the site due to extensive reshaping of the land. Contours must be restored to bring back that wetland, and temporary drainage may be needed to reshape the land. It may help to think of the need for further drainage as being similar to a physician having to re-break a partially-healed broken bone to set it properly, so it will heal well and have a better appearance.

This is how to determine if construction would be aided by digging ditches prior to building the wetland: if water seeps into test holes dug where soils will be needed to build the dam and groundwater dam, the project would benefit from further drainage prior to construction. If one fears becoming stuck while crossing the proposed worksite with a truck, the project would benefit from drainage. If one is uncertain of the need to drain a site prior to construction, invite an experienced contractor to visit it and ask his opinion.

Successfully draining a wetland construction site requires some planning. First, identify an outlet for the main ditch to be constructed, which is usually a creek or old ditch along the lowest edge of the project site. The main ditch to be dug will begin at this low place and extend uphill into the work area. Next, use a level to measure the elevation at this deep outlet, and where the dam will be built. The goal is to create a main ditch that will be three feet deep, and begins at the lowest place and extends uphill through the middle of the work area. For example, if the difference in elevation between the outlet and the upper end of the work area is five feet, a five-foot deep ditch can be started at the creek that will be three feet deep when it ends at the uphill edge of the work area.

A ditch should be built to have a slope of about one inch of rise for every ten feet of length. The main ditch should be placed to pass up through the deepest part of the wetland being built, with additional branches being dug to parallel the location of the dam. Branch ditches should also be dug to drain springs and areas of standing water. Overall, it is best to space these ditches about 60 feet apart through areas where soil is needed. One will find that during construction it will be possible to remove soils to a depth that equals the bottom of the ditch.

Digging deep ditches with vertical sides may cost less money, but these ditches will cave in before construction begins. Ditches should be benched or stepped back so they will not fall in. In wet areas, make drainage ditches three-bucket widths on top and one-bucket width in the bottom. A crew works best for digging these ditches. One person operates the backhoe or excavator, while the other uses the level to monitor elevations. An excavator and survey crew should be able to dig 200 feet of ditch per hour.

It is not possible to dig too many ditches prior to construction. Their effect on soil moisture is nothing short of amazing. Within a couple of weeks of digging, it will be possible to remove large quantities of soil in areas previously too soft to cross with heavy equipment. The cost of digging these ditches is more than made up by the quantities of soil that can be later moved and shaped to build the wetland.

Heavy equipment passing back and forth over a wetland construction site can mix in water already present in the ground and create mud, causing heavy equipment to become mired. It is a good idea to have two pieces of heavy equipment on the job site in case this happens. A stuck dozer can be pulled out with little loss in production when another dozer or excavator is available to help.

"Do everything you think of to get the wetland to hold water when you're at the construction site. When it holds water you'll feel lucky, because you can't think of anything else you could have done to make it work," says contractor John Utterback, who has built hundreds of successful wetlands.

When soils become wet, pay special attention to the construction process to ensure that work is being conducted correctly. The fear of becoming mired may cause the contractor to change construction techniques, resulting in failure of the wetland project. Monitor construction to ensure the contractor does not change how deep the groundwater dam is to be dug. Reducing the depth of the excavation may end in missing buried layers of sand, gravel, and drain lines, resulting in a defective wetland. When soils get wet, it may become necessary to dig a series of ditches for drainage, and return in a couple of weeks to complete the job. Since delay is not always acceptable, it may be best to pile soil for the dam and groundwater dam layers thicker than six inches, and push up a loose pile of soil for the dam. Since wet soils cannot be packed, the dozer operator can back blade the soils to shape a rough dam. These wet soils can be expected to settle one foot or more when dry. Keep the drainpipe open in these new wetlands, so they do not begin to fill. After two weeks or more, one should be able to return with a dozer to shape, pack, and finish the dam.

Water Control Structures

Installing a water control structure and drainpipe in a wetland provides flexibility in how the wetland can be managed. Being able to adjust water levels can be critical for establishing and maintaining plants, and for controlling unwanted plant and animal species. Managing water levels provides the opportunity to:

1. Expose mudflats for shorebird feeding during spring or fall migration.
2. Dry the dam in order to repair damage caused by muskrat and beaver burrows.
3. Eliminate carp that reduce aquatic plant diversity and abundance.
4. Dry saturated soils to reduce cattail density.
5. Remove predatory fish to improve amphibian breeding habitat.

Over 200 of the wetlands managed in Kentucky on the Daniel Boone National Forest are located within the flood pool of Cave Run Lake, a large flood control reservoir.

Occasionally, hurricanes elevate the lake level by 20 or more feet, inundating these wetlands. As the waters recede, the wetlands become stocked with carp, muskellunge, and bass. While this situation pleases those who like to fish, it spells disaster for waterfowl. Water control structures have been installed in these wetlands to manage the waters and the fish.

Fish can affect waterfowl by eating ducklings, the plants and invertebrates they require for food, and the plants they need to hide from predators. Use a dip net to sample a wetland containing fish, and one would be fortunate to catch a newt. Sample a wetland without fish and one will likely find salamander and frog larvae, dragonfly and damselfly larvae, fairy shrimp and crayfish, water boatman, and backswimmers, all waterfowl food.

While building a wetland, it is easy to think one would never use a water control structure, but one could be mistaken. In 2005, the author built a series of 20 wetlands in Kentucky that were intended to dry seasonally to provide breeding habitat for wood frogs and spotted salamanders. The maximum depth of each wetland constructed was only 18 inches in the center. However, the wetlands did not dry, even during a severe drought in 2007. They now contain high populations of red-spotted newts and bullfrogs, species that thrive in more permanent water, and which prey on the more desirable frogs and salamanders attempting to breed in the wetlands. Had drainpipes with water control structures been installed, these wetlands could have been periodically drained to maintain the conditions required by the less common amphibians.

The ability to remove water from a wetland is important when making repairs. When muskrats and beaver tunnel into dams, the large holes they create must be repaired before waters flow through them, causing failure. Repairing this damage involves opening up the burrow with a backhoe or excavator, and then packing it with soils high in clay. These burrows generally contain water, and are difficult to block unless the wetland is dry. Removing water from wetlands with pumps and siphons is expensive and takes time, while opening a water control structure can be done in minutes.

Cattails provide important hiding cover for juvenile wood ducks and mallards, yet a dense growth covering the entire wetland can reduce plant diversity and the attractiveness of a site. Fortunately, in many areas muskrats will take up residence and control the cattails by using them for food, and for building their houses. Another way of reducing cattails involves mowing the bottom of the wetland when it is dry. However, this is seldom possible unless there is a drought, or a water control structure has been installed. The water control structure can be opened in early summer, and after soils dry in the bottom, it can be mowed later that fall. The water control structure is closed after mowing, and the next year the density of cattails will be greatly reduced.

Shorebirds are some of the most enjoyable wildlife species to watch, often allowing people to come near them for identification. Species such as the avocet and ruddy turnstone migrate thousands of miles in the spring and fall. Shorebirds are experts at spotting mudflats when flying overhead, landing to feed in large flocks at wetlands showing an exposed perimeter of mud and shallow water. Unfortunately, some years mudflats are not present when shorebirds are migrating, forcing them to travel great distances without rest or food.

Water control structures allow managers to create conditions that are ideal for migrating and breeding shorebirds by raising or lowering water levels a couple of inches each day to expose mudflats independent of weather conditions. An added bonus is the creation of outstanding wildlife viewing opportunities. This practice also works on small wetlands in the mountains.

Newly-developed water control structures that are solar powered allow for the timed and remote adjustment of water levels in wetlands. These structures now permit individuals to manage water levels at a number of sites over a large geographic area, saving considerable labor and expense.

Building a forested wetland can be a challenge, as trees and shrubs will not germinate or grow in standing water and saturated soils. Therefore, it is necessary to maintain dry soils in locations where trees are desired within a wetland restoration project during the growing season. Installing a drainpipe and water control structure allows for adjusting water levels to maintain elevations needed to keep woody plants alive.

The water control structure and drainpipe should be located in the lowest portion of the wetland so that the majority of waters can be removed if needed. The basin of the wetland should be shaped so that waters will collect at the inlet of the drainpipe, not leaving pools of water that can cause problems by providing refuge for unwanted fish.

A disadvantage to installing a water control structure is that it adds something that will itself have to be maintained in the future. Wetlands can fail because water control structures develop leaks, can be vandalized, become damaged by flood, or may be broken when mowing. If one's goal is to restore a wetland that will require no maintenance in the future, it is best to design the wetland so it will meet objectives without installing a water control structure or drainpipe.

Perhaps one of the most troublesome items used as a water control structure is the flashboard riser, consisting of a vertical section of steel corrugated culvert that has been cut in half, with boards that are dropped down into a slot to raise or lower water levels. These structures are plagued with problems, and should be removed from wetlands whenever possible. First of all, they leak, so much so that many wetlands are unsuccessful simply because water leaves the wetland between the boards.

Secondly, it is practically impossible to keep the structures from leaking. Many wetland managers have labored for years to keep flashboard risers from leaking with limited success. These structures are also dangerous, and can result in the loss of human life. The danger occurs when they become blocked by beaver, and someone attempts to clear them. These structures can hold back a massive amount of water when plugged. Freeing them of debris requires hand labor, and places the worker in a position susceptible to drowning when the structure is finally opened.

Water control structures that are made of PVC that contain adjustable plates sealed by rubber gaskets work well for regulating water levels in wetlands. Products made by Agri-Drain Corporation and Fiberglass Utility Supplies are easy to install, maintain, and use. These structures are installed in the dam and draw water from the bottom of the wetland, which makes them less likely to be blocked by beaver. They also allow managers to adjust water levels without having to get their feet wet.

This inexpensive and adjustable water control structure was fashioned from PVC Schedule 35, SDR drainpipe with rubber gaskets.

Sources for Water Control Structures:

Agri-Drain Corporation
1462 340th Street
Adair, IA 50002-8578
(641) 742-5211
http://www.agridrain.com

Fiberglass Utility Supplies
1465 250th Street
Libertyville, IA 52567-8523
(641) 693-3311
http://www.dosir.com

Drainpipes

Schedule 35 PVC drainpipe with rubber gasket joints works better than steel or cement for drainpipes for a number of reasons:

- Connections between pipe sections are easy to make and will not leak.
- Water control structures are easily attached, and their connections will not leak.
- Connections between pipes are flexible, prolonging life under freeze-thaw conditions.
- Rust-outs will not occur.
- Repairs and adjustments can be made by one person.
- Drainpipe inlets can be beaver-proofed with success.

Schedule 40 PVC pipe is heavier and stronger than Schedule 35 PVC pipe, and may be required by

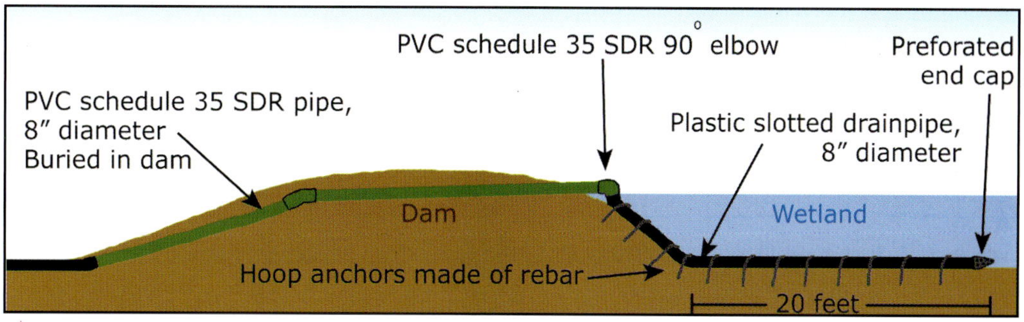

Overflow pipe modified to reduce plugging by beaver

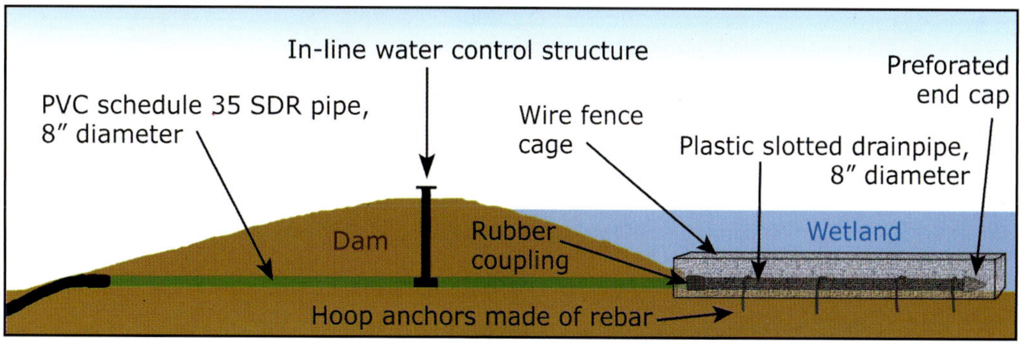

Drainpipe modified to reduce plugging by beaver

Beaver attempted to block this drainage ditch. Water continues to flow down the ditch by passing under the dam.

government agencies for projects they help finance. Schedule 40 is generally twice as expensive as Schedule 35, and is seldom worth the added cost for wetland projects.

To reduce costs on small wetlands with small watersheds one can forego the expense of installing a commercial water control structure by simply attaching a threaded clean-out trap to one end of the pipe so that waters can be drained.

A simple and inexpensive water control structure can be made by attaching a 90-degree elbow with a vertical section of pipe to the end of a drainpipe. The top of the pipe should be cut at the desired elevation of the water level. These structures should be attached to the outlet of the drainpipe in order to reduce possible problems with beaver blockage. They work best in shallow wetlands on more level ground where the vertical pipe can be kept under four feet high. A primary disadvantage is that they can be easily vandalized.

It is always a good idea to modify the inlet of the drainpipe to reduce beaver problems when installing a water control structure. Beaver can detect water flowing through the drainpipe and may block its entrance with debris. The inlet pipe should be extended into the wetland by attaching from 20 to 30 feet of single-wall, slotted, or perforated plastic corrugated drain line to the entrance of the drainpipe. This extension will diffuse the velocity and sound of water flowing into the pipe when the wetland is drained, lessening the chance of beaver detection. The pipe extension should be anchored to the bottom of the wetland, and sections of rebar bent over at the top work well for this purpose. Building a wire cage around the inlet pipe helps even more.

Blocking Ditches

Many still believe that in order to restore a wetland, one only needs to plug a ditch with soil. Unfortunately, such action does little to bring back the original wetland. Ditches are often the only visible feature that remains after extensive work has been done to drain a wetland. Fields were almost always reshaped so they would drain, initially by using livestock, and later by heavy equipment, to direct runoff into constructed ditches. Farmers used animal-drawn scrapers, and later graders and dozers, to fill low places, and then excavated wide, naturally-appearing pathways for water to follow into the open ditches.

What is seen on the surface of a ditch provides small indication of what has been done below ground to drain a wetland. Drain lines made from clay, wood, rock, and plastic were often buried in the bottom of the ditch, or alongside the ditch to keep it dry enough to cross with farming equipment.

Restoring a wetland drained by a ditch should involve construction of both below and above ground dams across the lower edge of the area traversed by the ditch, and reshaping of the lands that form the watershed for the ditch to return a depression where the wetland used to be.

The Construction of Groundwater Wetlands

Chapter 5

A groundwater wetland is supplied primarily by water that is found underground. It is generally built by removing soils to create a depression that exposes the water table, making the water in the wetland the same elevation as the water table. It may help to think of a groundwater wetland as resembling a large diameter open well.

The texture of soils surrounding a groundwater wetland can vary from sand, gravel, and clay, to silt and peat. These soils are usually saturated, permeable, and not compacted. While the elevation of water in most groundwater wetlands remains stable, some may change as much as six feet over the course of a year. Water levels in these wetlands will vary as the water table rises in the spring and drops in the fall.

Groundwater wetlands are simpler to build, in some respects, than surface water wetlands, because soils do not have to be high in clay or compacted during construction. However, there can be problems with them going dry, should the water table drop during drought or pumping from the aquifer.

Some areas where groundwater wetlands can be built may already be wetlands. Because the water table is high, the site may be growing aquatic plants, and standing water may be present during part of the year. If it is determined that a wetland construction site is already wetland, one may be able to help the environment more by changing locations to a site that is not already a wetland. Before proceeding with the construction of a wetland on top of an existing wetland, it should be determined if the project being considered will be of greater value than the wetland already present on the site. After making this decision, required permits must be obtained before proceeding.

This groundwater wetland was built with an excavator in Menifee County, Kentucky.

There are many reasons why someone may want to change all or part of an existing wetland from one type to another. Often, the existing wetland is not the same that was present on the site historically. The existing wetland could be partially drained, with some of the historic actions taken to eliminate it now healed, returning wetland characteristics to the site. Modifying the wetland may be a response to specific needs, such as endangered species habitat, shorebird migration habitat, amphibian breeding, fish habitat, wildlife viewing, esthetics, flood control, water purification, or environmental education.

Wetlands, like all other ecosystems, change over time, and this change can be rapid. For example, beaver may be responsible for altering wetlands from one type to another. Rivers and streams often adjust their courses during major flood events, creating and destroying wetlands in the process. People in coastal areas are often reminded of the way hurricanes can make wide-scale, overnight changes in wetland diversity and abundance over hundreds of thousands of acres. The government allows wetlands to be modified from one type to another, providing required approvals are obtained in advance from permitting agencies.

Determining How Deep to Dig

Because it is possible that the elevation of the water table may vary with the season, it is important to know how deep groundwater is below the surface during the driest time of year. A simple posthole digger or soil auger can be used to measure the elevation of water in the ground. Simply dig a test hole in the center of the proposed wetland location. After waiting a few minutes, water should seep into the hole from the bottom and sides, stabilizing at an elevation equal to that of the water table. If the water rises to one foot below the surface, and the desired maximum depth of the wetland is two feet, a depression at least three feet deep must be excavated. It is important to keep in mind the time of year the test hole is dug, as the water table can be expected to be highest in early spring and lowest in early fall. If one is designing a permanent water wetland, it would be best to dig the test hole during the driest time of year. However, if one's intent is to provide breeding habitat for amphibians dependent on ephemeral wetlands, it would be best to dig the test hole when larvae would be metamorphosing.

Water exposed beneath the root mass of a fallen tree indicates a high water table.

Groundwater in a test hole dug by a backhoe shows the elevation of the water table is over 6 feet below the surface.

Construction Steps

Hire a skilled operator and excavator to build the wetland. The excavator should be a 100-Series or larger. At a reasonable cost an excavator can be used to create a naturally-appearing wetland that exposes groundwater in almost any location that has a high water table. Excavators excel at digging in soft areas where dozers become mired. The wide pads on an excavator allow it to travel in places where a backhoe has trouble moving. An experienced operator can create gradual slopes, feather out excess soil, and place woody debris with the skill of a sculptor. The excavator allows a wetland to be shaped so that it looks like it has always been there. A skilled operator should be able to build a wetland measuring 30 feet in diameter and two feet deep with an excavator in half a day.

To construct a naturally-appearing wetland that is 30 feet in diameter, an opening in the forest that is at least 60 feet in diameter must be made. Trees get in the way of the swinging boom of the excavator, so it is important to make room available for the machine to work. Space is also needed to spread the soils removed to form the basin. As it is much faster for the excavator to topple trees than it is to remove stumps, there is no need to cut trees with a chainsaw prior to construction. The wetland will also function better and appear more natural if entire trees with roots attached are placed in and around it following construction.

Not all trees have to be removed in the area marked for the wetland. Some trees can be left standing within and along the edge of the work area, providing there is enough room for the excavator to maneuver. Even though working around a few trees may take longer and cost a bit more, the action will help the appearance and maintain shade over the wetland, if desired.

This small groundwater wetland was created by modifying a drainage ditch at Nancy Elementary School in Kentucky.

Begin construction by showing the operator an area encircled by brightly-colored flags, and ask that the trees and shrubs within the flags be cleared. Instruct the operator to place trees and shrubs that are removed outside the flagging in small piles, so they can be later placed in the finished wetland, or left hidden so they will not obstruct the view.

Use wire flags to mark the perimeter of the wetland after trees and shrubs have been removed. Set up a construction level out of the way on high ground overlooking the construction site, and record elevations on the surface along the wire flags. Elevations should change no more than two feet from upper to lower edge. Adjust flags so the perimeter is marked on the flattest ground possible, leaving room between the flags and trees along the lower edge for spreading excess soil. Then, determine the lowest elevation along the flags to use as a reference of the maximum depth while building the wetland.

Use the excavator to remove vegetation and topsoil within the area marked by flags, and place the material in piles along the edge. Construct a groundwater dam around the lower two-thirds of the cleared area. This involves digging a trench at least 12 feet wide that cuts through tree roots, intercepts buried drainage structures, and extends below the bottom of crayfish holes. Return the soils that were removed to create the trench, and compact in layers using the excavator bucket and tracks. It is not possible to obtain complete compaction of soils that are saturated or low in clay.

Dig a hole in the center of the wetland that will be the maximum depth in relation to the elevation recorded along the lower edge of the marked perimeter. The maximum depth is equal to (elevation of water table below surface) + (maximum depth of water desired) + (thickness of vegetation and topsoil to be placed back in depression).

Once the maximum depth is established, extend the

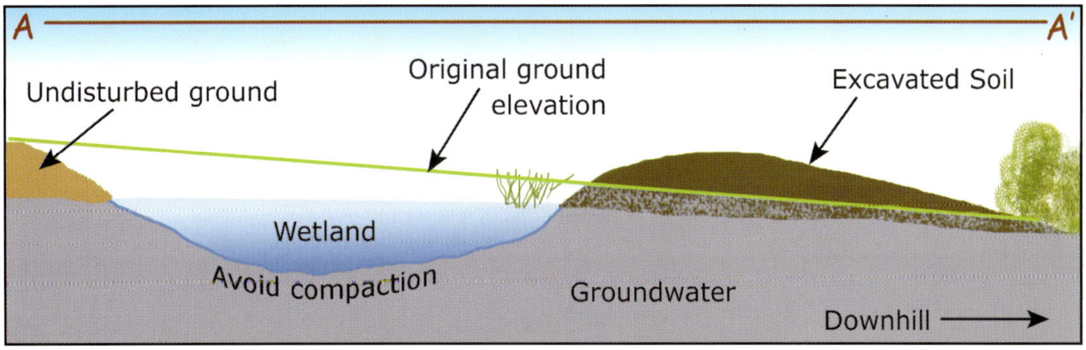

Construction of a groundwater wetland, Profile view

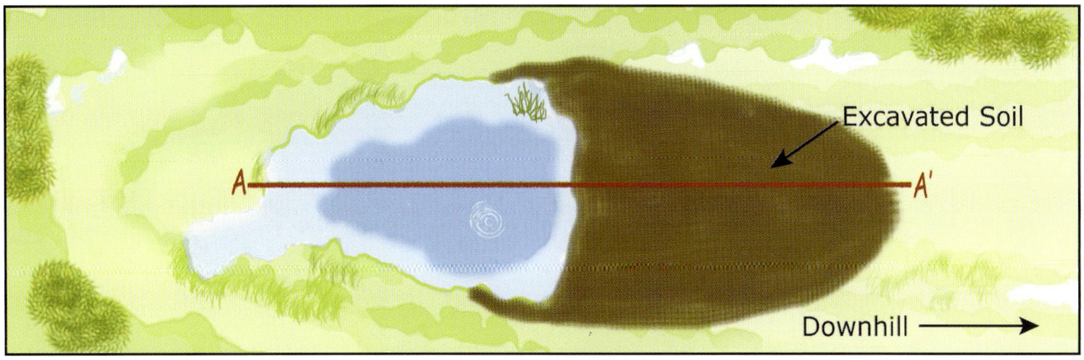

Construction of a groundwater wetland, Plan view

bucket of the excavator toward the center of the area, and begin drawing soil up to the machine to create a depression. The depression should be deepest in the middle, and then come uphill to the flags with a gradual slope. The excavator rotates after each scoop to deposit soil between the machine and the edge of the clearing, and then moves over for another set, digs inward again, and continues the process until the depression is complete. Check the depth of the depression often to ensure that it is acceptable. If needed, the operator can set up and dig another round, making it deeper, and adjust slopes.

It is important to cut through and remove vegetation and topsoil around the lower edge of the depression before excess soils are spread over the area. Otherwise, waters can leave the wetland by following the porous layer of vegetation and topsoil buried beneath the piled soils.

Soils removed to create the depression can be spread down slope to create a dam. The dam will trap additional runoff and increase the size and depth of the wetland. While the dam and groundwater dam may not contain waters as well as one constructed of compacted clays, it can be expected to significantly slow the loss of surface runoff from the wetland.

Once the depression is complete, compacted areas in the bottom are loosened so that groundwater will seep into the new wetland. The excavator then spreads the saved vegetation and topsoil over and around the depression. Excess soils are spread outside the depression along the lower edge and blended with the surrounding landscapc. Avoid piling soil uphill from the depression that can wash into the wetland or prevent runoff from entering the wetland.

Groundwater wetland construction progression: Small wetland using 100-Series Excavator

Orange wire flags mark the perimeter of the wetland to be restored on a site with a high water table.

The excavator removes vegetation and topsoil, piling it outside the flags.

Vegetation and topsoil are cleared from the entire worksite.

A wide trench is dug along the lower two-thirds of the cleared area. The trench extends below the bottom of crayfish burrows and intercepts buried drainage structures.

The trench is filled with mineral soils, with deep layers being compacted by using the bucket on the excavator.

The upper layers of soil placed in the trench are compacted by traveling over them with the excavator.

Soils are added to the trench in layers and compacted by traveling over each layer four times with the excavator. Saturated soils are compacted by only using the bucket.

A depression is dug in the center with a depth = (elevation of water table below surface) + (maximum depth of water desired) + (thickness of vegetation and topsoil to be placed back into depression).

Gradual slopes are placed around the depression, with clay soils being used to build up the lower edge.

Excess soils are spread for a considerable distance from the lower edge to make them blend with the surrounding area.

Excess soils are blended into the adjoining slope. They are not compacted to provide turtle nesting habitat.

The bottom of the depression is loosened so groundwater can seep into the wetland.

The completed wetland was seeded to wheat, and mulched with leaves and straw.

Here is the same wetland three months after construction.

Groundwater wetland construction progression: Large wetland using 200-Series Excavator

The excavator begins by removing and saving vegetation and topsoil.

Vegetation and topsoil are piled along the outside edge of the colored flags.

A hole is dug in the center with a depth = elevation of water table below surface) + (maximum depth of water desired) + (thickness of vegetation and topsoil to be placed back into depression).

The mineral soil removed to create the depression is piled in back of the topsoil. The large 200-Series Excavator has a long reach, allowing for efficient sorting of soils.

Gradual slopes, beginning at the colored flags, are placed into the center of the wetland. Groundwater seeps into the deepest part of the wetland.

Topsoil and vegetation are spread over the depression.

Slopes are covered with the topsoil and vegetation that was saved.

Piles of mineral soil are spread so they blend into the landscape.

The majority of mineral soil removed is spread downhill, so it will not keep runoff from entering the new wetland.

Large trees are placed in and around the wetland to control vehicle access.

Exposed soils are seeded to wheat and mulched with wheat straw to control erosion.

The new wetland will fill as the water table rises over winter and spring.

Care must be taken when building on organic soils lest the soils erode into the wetland over a couple of years. To prevent sloughing, it is necessary to maintain slopes ten or less percent surrounding the excavation. A large area must be cleared on peat to create a deep wetland with gradual slopes. For example, a cleared area 130 feet or greater in diameter is needed to build a wetland six feet deep in the center with gradual surrounding slopes.

This is an example of a newly-constructed emergent wetland built from organic soils on a site with a high water table. Gradual slopes combined with spreading soils far away from the depression will help prevent sloughing.

An excavator shapes soils where it would be too soft for a backhoe or dozer to work.

Perhaps the most efficient way of constructing a groundwater wetland with an excavator is to pair the machine with a dozer. The dozer, in most cases, will do a faster job of leveling the soils removed by the excavator. These soils can be pushed outward from the edge of the clearing between larger trees, like spokes on a wheel, to further improve the shape and appearance of the site.

This dozer spreads soils removed by the excavator in building a groundwater wetland.

Skilled operator Billy Osborne (shown here) builds a groundwater wetland on mud where other dozer operators would find themselves stuck.

Groundwater wetland construction progression: Large wetland using scrapers

The lowest portion of a large field is chosen for building an ephemeral wetland above the Trinity River in California.

A test hole dug by an excavator shows gravel soils, with groundwater 8.5 feet below the surface.

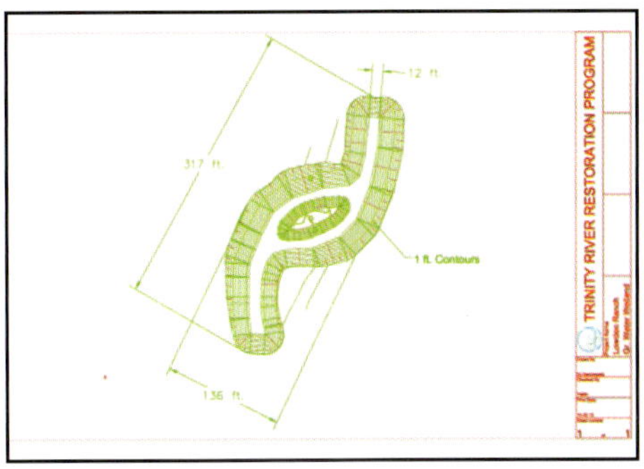

The design for the wetland prepared by David J. (DJ) Bandrowski, P.E., incorporates gradual slopes, an island, and a shape that can be efficiently constructed by using scrapers.

DJ (yellow vest) works with personnel to mark the location of the wetland to be built.

DJ explains the wetland project to the construction manager.

The scrapers follow the flagged perimeter to begin the excavation by removing topsoil.

Two scrapers, which are linked together, push and pull each other to remove over 30 cubic yards of soil with each pass.

The excavator creates a 30-percent slope on the side of the basin.

Soils are placed by the excavator in front of the scrapers for removal.

A laser level and rod are used to check elevation in the bottom of the depression.

Groundwater seeps into the bottom of the depression.

Large dump trucks are used to remove soils when the bottom of the depression becomes too soft for scrapers to operate.

Large woody debris is placed in the depression.

A layer of topsoil is spread over the gravel.

Boulders and slash have been placed in the depression. Approximately 5,000 cubic yards of soil was moved to create this one-acre wetland.

Wheat is sown for erosion control. Note how the logs are placed for possible use by the western pond turtle.

Wheat straw is spread for mulch.

Large piles of slash are placed in the field around the new wetland to give amphibians and reptiles using the wetland a place to hide.

Here the wetland has filled with groundwater the following spring. (USBR Trinity River Restoration Program photo)

Black Diamond Wetland Restoration Project

Sarma Liepins, an ecosystem manager for the British Columbia Ministry of Environment, uses excavators to restore groundwater wetlands in the semi-desert region of BC near Kamloops. She regularly walks her dogs in an area called Tree Flats, a rolling savannah of ponderosa pine, where many trees have been turning brown after losing their battle with the pine beetle. The open grassland is dotted with scattered depressions, each forming a shallow wetland from 60 to 90 feet across. A few of the wetlands contain shallow water fringed with sedges and bulrushes, while most are muddy or dry. White powder from historic, volcanic ash deposits forms a ring around these wetlands. The waters in these wetlands is clouded with suspended soil, and choked by thick mats of algae. Evidence of cattle trampling is everywhere, and the surrounding upland is as unyielding and solidified as a parking lot. OHVs are using the wetlands for a playground, with their tracks covering the muddy wetland basins.

Cattle have compacted clay soils in this natural ephemeral wetland near Kamloops, British Columbia.

Over the years, Liepins has monitored amphibian populations in what she and friends affectionately called the Black Diamonds. She found that while spadefoot and western toads breed in these habitats, they rarely hold water long enough for tadpoles to develop. The wetlands may actually be harming toads by luring them to breed in places where they cannot live to adulthood.

The author first viewed the wetlands with Liepins in 2006 and questioned how they were being supplied with water. Soils were sampled in the bottom of three of the wetlands using a hand-held auger, and they were found to be high in clay, as three-inch long ribbons formed. It was assumed that these were surface water wetlands where their waters were being maintained by snowmelt and runoff.

In June 2007, Liepins and the author looked for ways to direct more runoff into the drying basins to lengthen the time they could hold water. Several roadside ditches were found to be diverting waters from the wetlands, but redirecting these did not seem to be the long-term, drastic action required to change how long waters would remain in the wetlands.

Observing damage caused by an OHV showed how these wetlands historically functioned. A 4WD mud bogger had become stuck in the bottom of one of the dry wetlands, so a backhoe was brought in to dig out the mired vehicle. The episode left an open trench in the bottom of the wetland approximately 12 feet long and three feet deep. The trench was filled with water, while the rest of the wetland remained dry.

Apparently, the backhoe had dug through a layer of packed clays in the bottom of the wetland, and released groundwater trapped beneath. Waters were now rising to the surface in the bottom of an otherwise dry depression. Historically, these wetlands had been maintained by a combination of surface water and groundwater. However over time, unrestricted livestock access combined with OHV traffic had packed the bottom of the wetlands, and this impermeable layer of clay was now preventing groundwater from reaching the surface.

To test this theory, holes were drilled using a hand auger in the bottom of three adjacent wetlands. The top 12 inches of each were very difficult to penetrate, showing significant compaction. Soils in each test hole loosened considerably once the auger reached the 12-inch mark, with waters seeping into auger holes below the compacted surface layer, and quickly topping near the surface.

One of the main techniques used in restoring wetlands involves locating an area with soils high in clay, and then packing these soils so that they hold runoff. While the Black Diamond wetlands were holding runoff, their compacted soils were preventing groundwater from contributing to the wetland water supply. The compacted clays in the bottom were acting like the lid on a container of water, preventing groundwater from rising to the surface.

Rancher Ray Frolek described a similar problem he observed with constructed ranch ponds in British Columbia that would not hold water. Frolek had repaired ponds that no longer held water by breaking up compacted soils in the bottom. He found this problem occurred when cattle were allowed complete access to a groundwater-based pond that was constructed from clay soils. The cattle's tromping compacted and sealed the bottom, preventing ground-water from filling the pond.[40]

Cattle also changed the shape of the Black Diamond wetlands. They now have a flat bottom with steep, almost vertical sides. Examining the topography surrounding each wetland shows they once had gradual slopes, with the deepest water found in the middle. Again, cattle modified the wetlands by reducing their depth, and undoubtedly, reducing plant diversity by giving the wetlands a uniform, shallow depth.

A plan was developed for restoring the Black Diamond wetlands that involved loosening their soils, reshaping their basins, and protecting them from further damage. An excavator was used to remove and save the vegetation and topsoil in each wetland, temporarily piling this material around each wetland. The excavator reshaped the wetland basin, making it deeper in the bottom with gradual slopes. Topsoil and vegetation were then spread back over the bottom of the wetland. The operator took care during the project not to pack soils with the excavator bucket or tracks. After the topsoil was spread, a number of large trees were laid in and around the wetland to form a natural barrier that cattle and OHVs were unlikely to attempt to cross—not more than once, anyway. Exposed soils were seeded to winter wheat and mulched with straw to reduce erosion.

It is believed that the historic compaction of soils high in clay within wetlands supplied primarily by groundwater is widespread throughout western North America.

Blasting Method

Explosives are an effective way to build small wetlands where heavy equipment cannot be used, and where vehicle access is difficult or not advised. Blasting can be used to establish small wetlands of various depths, at any time of the year, and at a reasonable cost. The blasting agent consisting of a mixture of ammonium nitrate and fuel oil (AN/FO) has been used since 1963 to establish thousands of wetlands in North America.[41]

Blasting should not be considered if the potential worksite is within 2,000 feet of any home, business, or structure such as a bridge, tower, pipeline, or transmission line. Potential for damage from vibration, flying rock, vegetation, and soil may be too great within this zone. Blasting that is done for activities such as road construction within 2,000 feet of these improvements is done under carefully controlled situations where pre- and post-blasting surveys are conducted by qualified personnel to protect the blaster against fraudulent liability claims.

By law, a licensed blaster is required for projects that involve the use of explosives. The licensed blaster is the only individual who can legally purchase, transport, store, and use the blasting agents for a project. One can find blasting expertise by looking in the *Yellow Pages* under "Explosives". A company that markets explosives will be able to direct one to a licensed blaster in the area. One may also wish to contact the International Society of Explosive Engineers for information.

It is a good idea to invite a licensed blaster to visit potential worksites to help plan a project. Completing these steps beforehand can make the most out of the blaster's visit:

- Select an area 2,000 feet or more from where people live, and from improvements.
- Test the elevation of the water table, and determine how deep the excavation should be.
- Mark the desired perimeter of each wetland pool area with flagging.

Explosives are best used to create small and shallow wetlands. It is difficult to predict the actual size and depth of the wetland to be created with a given amount of explosive, due to variations in soil texture, depth to bedrock, water content, and type of explosive used. Best results are obtained on sites that have a firm base, and not on floating bogs. On the Daniel Boone National Forest, a typical charge of 100 pounds of AN/FO was used to blast wetlands 30 feet in diameter and three feet deep in silt-loam soils, which, in turn, rested on five feet of gravel that overlaid bedrock. It appears realistic to expect a similar charge to create wetlands from three to five feet deep, ranging from 22 to 41 feet in diameter on other locations.[42]

When blasting, the depth of the recently-created depression will appear much greater than the eventual water depth in the new wetland. Displaced soils create a raised rim around the wetland, and this rim will not maintain water higher than the elevation of the water table. Soft soils around the edges can also be expected to slough, filling the wetland even more.

Explosives were used to create this wetland on the Monongahela National Forest in West Virginia over 40 years ago.

AN/FO is generally the lowest-priced explosive to use for blasting wetlands, but it has its disadvantages. The material must be kept dry, which can be difficult when water is standing at or near the surface, and gases produced from the blast are poisonous, so one must take care not to be exposed to the fumes.

Usually, when AN/FO is used, holes are dug at a 12- to 15 -foot spacing by using a shovel or posthole digger. AN/FO, which is kept dry in plastic bags, is placed in the bottom of the hole, and then the top of the hole is filled with soil and packed well with an iron bar. The blaster will have inserted a blasting cap into each bag of AN/FO for detonation. When using multiple bags of AN/FO, delays must be used in order to have time to move soil away from the site to prevent it from falling back into the hole.

Licensed blasters generally prefer to use more modern explosives when creating wetlands. Certified USDA Forest Service Blaster Jon Hakala, on the Superior National Forest in Minnesota, prefers to use an emulsion instead of AN/FO as a blasting agent. An emulsion has many advantages over AN/FO: it is lighter to carry, does not have to be kept dry, and its placement involves much less digging, as a hand-held auger can be used to quickly prepare the holes. Another advantage is that an emulsion will not leave behind a residue of ammonia and petroleum, which are considered pollutants.

Hakala suggests purchasing packages of blasting agent that weigh four pounds each, which come in sausage-shaped containers measuring 3.5 inches in diameter by 24 inches long. Three of the sausages should be tied together, and then each set sunk in separate holes dug in a circular pattern to make a round pond, or staggered to make an oval pond. The holes for the blasting agent can be kept from collapsing by placing a plastic pipe in them as soon as they are dug. A long pole can be used to push the explosive down the center of the pipe to the bottom of the hole. The plastic pipe can then be removed for use in other holes.

Spring Development

Ponds have been created from springs for hundreds of years. A spring is simply a place where water flows to the surface from within the ground. Springs are generally located at or near the base of a hill. It is rare that water can be seen coming out of the ground like a water fountain, as the flow usually seeps into soils, creating mud that is trampled by deer and hidden by vegetation.

There are a number of clues that show the location of a spring: Waters flowing from a spring will usually form a small stream. Springs can often be found by following a stream uphill until it ends in a wetland. Springs are often surrounded by lush growth of ferns and sedges. The actual emergence of the spring can be an area of black-colored soil and melted snow. The spring contains soft, saturated soil surrounded by dry, firmer ground. The area where the spring emerges can be as small as three feet across or larger than an acre.

To find the source of the spring, one should walk uphill until one reaches the uppermost extent of soft ground. Dig a test hole, and water should flow into the hole from all sides, including the bottom. The color of soil should be gray or black.

The source of a spring may be visible on a topographic map or aerial photograph as a valley that begins on a side slope near the base of a mountain. Generally, the valley flowing down from the spring will look like someone started digging a ditch into the side of a hill. Often, larger trees will be growing around the source of the spring, as these soggy places required frequent hand-cutting by farmers to keep them open.

Under natural conditions, wetlands should be found around the spring and on the area down slope from the spring. Wet-meadow, ephemeral, and emergent wetlands are commonly formed by waters from a spring. In hill country, natural spring-fed wetlands were often long, winding, and narrow, extending downhill to a stream or river. Because a spring can saturate a large area of land, great efforts were taken by farmers to confine them and to direct their waters in order to dry wetlands for farming.

A small patch of wet ground, four feet across, at the base of a hill shows where a spring emerges whose flow was placed underground in six-inch diameter clay drain tiles leading to a channeled creek on the Wayne National Forest in Ohio.

The source of the spring generally offers opportunities for restoration, as a majority of these sites have been trampled by livestock. A natural spring should emerge from mud and be surrounded by gradual slopes; however, cattle will compact soils over the source of the spring, and flatten it so that it is level. These compacted soils can reduce flow from the spring. What is especially noticeable is how livestock shear edges around the spring, creating almost vertical banks. Wintering habitat for turtles, frogs, toads, and salamanders can be improved by taking action to reverse compacted soils and steep slopes that may surround a spring.

Wetlands of various depths can be created from springs by excavating a depression over or immediately downhill from the spring. Spring-fed ponds that supplied livestock with water have been dug by hand and with the help of horses, mules, and oxen since the 1700s. The technique involved digging out the hillside into the source of the spring, creating a depression that held a pond of water, and then building a rock wall in front of the cut slopes to keep soils from sliding into the pond. Many of these spring-fed wetlands are still present today; however, others have since been modified into larger ponds using heavy equipment.

An open water wetland can be formed from a spring by using an excavator or dozer to create a depression over

A dozer was used to build this emergent wetland fed by a spring flowing from the base of the hill.

Black leaves, dark saturated soil, green sedges, and water on the surface mark the location of the spring supplying water to this emergent wetland.

A stream flows from the spillway of this emergent wetland.

the top of the spring. The depth of water in the new wetland will be about the same as the depth of the hole that is dug over the spring. Soils removed from the depression are spread evenly downhill from the excavated area to create a naturally-appearing wetland, eliminating the appearance of a dam.

Spring restoration progression:

Sedges mark the emergence of a spring whose overflow was placed in a buried drain line and shallow ditch on the Roosevelt National Forest in Colorado. The backhoe operator removes and saves vegetation covering the compacted spring.

Soils are removed to create a shallow basin and excess soils are moved to a log landing and spread over non-native plants.

Gradual slopes are created around the depression.

Aquatics are replanted on slopes around the edge of the new wetland.

A groundwater dam is created across the ditch below the depression to block the buried drain line.

The wetland is finished the same day of restoration.

Wetlands built from springs usually contain water the entire year, regardless of their depth. Excess waters can be expected to flow from spring-fed wetlands in every season. This regular overflow offers excellent stream restoration opportunities. Therefore, where possible, the spillway leading from a spring-fed wetland should be fashioned into a naturally-appearing stream, instead of a ditch, to further improve habitat for plants and animals.

Historic ponds created from springs offer excellent wetland restoration opportunities. The excavated soils may be reshaped into shallow basins to create a series of wet-meadow, ephemeral, and emergent wetlands, all supplied with waters from the original spring.

Recharging Groundwater

Wetlands can be designed to collect runoff from the

surface and return it to the ground. This may be important in regions where the elevation of the water table is dropping due to water removal for irrigation or urban development. The role of wetlands in replenishing the aquifer has received much attention in the media, and is often a reason given for protecting wetlands.

Building a wetland that will actually inject water into the ground requires searching for a location that has a seasonally high water table, meaning the water table is near the surface during part of the year, and then drops lower when the seasons are drier. The fluctuating water table provides conditions where a wetland can be built that will develop the hydrology needed to support aquatic plants and animals, and will allow surface water to soak into the ground.

Measure the elevation of the water table by using a soil auger or a posthole digger to dig test holes over possible locations during two seasons, once during the wet time of year, and again in the dry time of year. In all probability, the soil texture on a suitable site will feel gritty because of its sand content, and may even contain visible layers of gravel. Chances are it will not be possible to form a two-inch long ribbon from the soil when placed between one's thumb and forefinger, but that is acceptable for this purpose.

The building of the wetland is, in many ways, similar to the construction of a surface water wetland, in that it is also accomplished by excavating a basin and building a dam to capture runoff. The main difference is in how the groundwater dam is constructed. When using this technique, the main purpose of the groundwater dam is to block buried drainage structures that can rapidly remove water from the wetland, but not the slower movement of water in more permeable soils. Buried plastic drain lines, clay pipes, wood boxes, and rocks can defeat the purpose of wetlands designed to recharge groundwater by quickly moving surface waters under the dam to empty into a ditch, creek, or river. The groundwater dam will block these structures. Drainage structures should be severed and blocked with fine-textured soils when found during construction. Often, a layer of fine-textured soil can be found over the more permeable soils within the planned wetland area. These soils can be scraped off and used to block drainage structures that are found in the trench being dug for the groundwater dam.

During construction, watch for the presence of soils which are high in clay. Should they form a layer over the bottom of the wetland, they could become compacted by heavy machinery. These compacted soils may form a barrier to water movement, and convert the site into a surface water wetland. However, this will not happen in areas where burrowing crayfish are present, as their holes provide a direct route for surface water to enter the ground. If there is concern about the bottom of the wetland holding water, patches of clay can be scraped away to expose the more permeable soils beneath, or their surface can be broken by pulling a deep sub-soil ripper behind a piece of heavy equipment to restore permeability.

The opportunity exists with these projects to capture large amounts of runoff by building wetlands with great surface areas that also have higher dams to maximize the quantity of water captured and added to the ground. Wetlands constructed in this manner can be expected to fill during the wettest time of year, and will probably dry as temperatures warm and rainfall decreases. The water levels of these groundwater recharge wetlands can be expected to fluctuate more than in surface water or groundwater wetlands.

Building a dam and creating a depression on a site that does not have a seasonally high water table is another way to capture runoff and add it to the ground. However, one cannot expect a wetland to form on such a site. The soils are generally so porous in such a location that waters will not puddle, and the soils will not become saturated, which means that conditions will not be suitable for wetland plants and animals. These dry depressions may both look and function like failed wetlands. However, it is possible that, over time, small wetlands will form in the bottom of such sites as fine-textured soils collect that will slow the downward movement of water.

Storm Water Wetlands

Storm water retention ponds in most communities are unattractive, expensive structures that provide little benefit to the environment. They do not have to be that way. For considerably less expense, these ponds can be built to appear and function like wetlands, and add beauty to the neighborhoods they are designed to serve.

A storm water pond should function by collecting runoff, and then allowing it to soak into the ground. It should be built on soils that are low in clay, not compacted, and where there is not a high water table. These construction factors allow the waters it collects to be filtered, reduce flooding, and replenish groundwater. A diversity of plants will grow on sites that contain water infrequently, and they also help clean and remove water.

Deep testing of soils should be completed when designing a wetland for treating storm water. When building in urban areas, it is often possible to simply loosen compacted soils already in place instead of excavating, removing, and replacing with purchased soils. In other places, it may be possible to break up an impermeable rock layer beneath the proposed site by using an excavator with a ram attachment.

Most storm water retention ponds are built with elaborate cement or steel water control structures that are designed to release waters at a lower rate than they entered, and that can handle overflow under flood conditions. Unfortunately, waters flow through these structures at such a rapid rate, they generally release the sediment and pollutants that entered them, allowing for the continued contamination of waters downstream. To prevent this, it is important that storm water ponds retain water long enough for it to evaporate or soak into the ground.

Storm water ponds built with steep dams often require a great deal of maintenance. The dams are subject to damage by muskrats, woodchucks, tree roots, and washouts from flooding. Steep dams are also difficult and expensive to keep free of trees because they have to be trimmed by hand. Building these dams with slopes ten or less percent will reduce maintenance and greatly improve the appearance of the pond. The gradual slopes will provide conditions for greater plant diversity, which generally results in increased opportunities for wildlife viewing. Gradual slopes also do not need rock to protect them from erosion, and because there is little rock and no drop-offs, the feature is a safer place for everyone to enjoy.

Storm water ponds built on permeable soils may not need a water control structure. To protect the dam from flood damage, a wide vein of large rock can be placed in the dam and passing through it to serve as a spillway. The backside of the dam should be hardened with rock and soil. Vegetation growing in the soil will help protect the dam from washing should there be a major flood.

This unattractive storm water pond was built with a steep, high dam that looks like a doughnut.

It is recommended that small ephemeral wetlands be created in the bottom of any storm water basins that do not hold water. This improves the value of them for plants and animals. Make a shallow depression in the bottom of the pond and install a synthetic liner, and cover the liner with soil. The shallow wetlands will hold water long enough for frog, toad, salamander, and dragonfly larvae to develop, and will not reduce the design capacity of the structure. The resulting benefit to wildlife populations should be dramatic by creating an ephemeral wetland surrounded by a wet-meadow wetland, which is ideal habitat for many species of reptiles and amphibians.

Forested Storm Water Wetlands

Streams that flow infrequently can be excellent locations for building wetlands to remove soil from runoff on forested land. A series of small wetlands for treating runoff and increasing amphibian habitat were built on forested lands near Port Alberni on Vancouver Island in September, 2008. The landscape surrounding the work areas was dominated by well-drained mountainsides of mixed gravels and rock, with scattered zones containing a water table high enough for wetland construction.

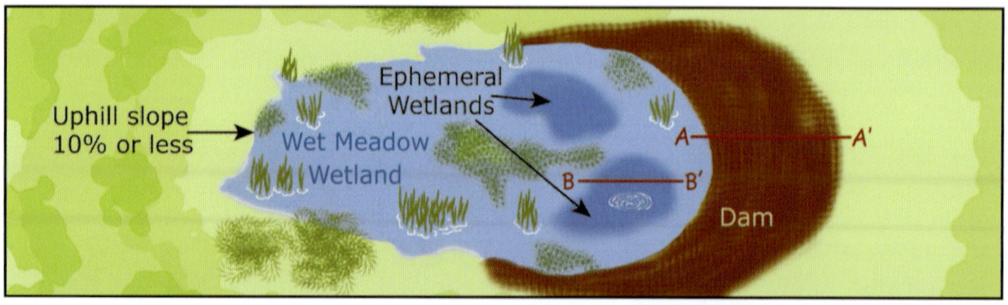

Storm water pond with ephemeral wetlands, Plan view

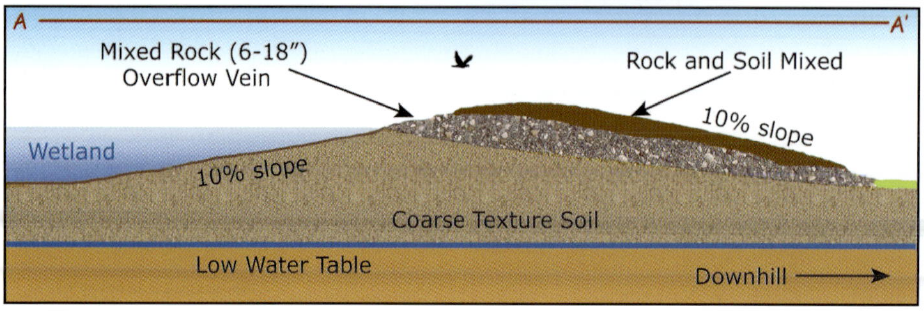

Storm water pond with permeable rock overflows

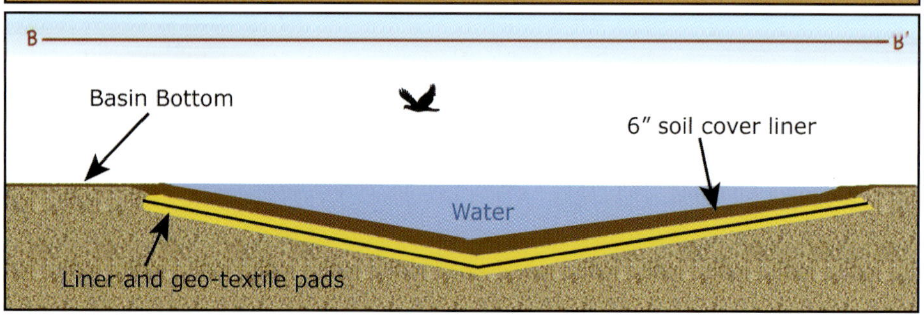

Ephemeral wetland constructed in bottom of storm water pond, Profile view

Ephemeral streams were located within timber sale units that were harvested from three to six years ago. Flowing down slopes ranging from 0.5 to three percent, the bottoms of the streams were vegetated primarily by sedges. Fireweed, sword fern, bracken fern, raspberry, and planted fir trees grew along the banks.

To identify wetland sites, ephemeral streams were walked to locate wet areas with exposed mud and patches of sedges and rushes. These mudflats were generally small, ranging from six to 30 feet long, and from nine to 12 feet across. Skunk cabbage, cattails, sedges, and bulrushes typically grew around the mudflats.

Soil texture and the elevation of the water table were determined for each mudflat by using a four-foot steel tile probe. The elevation of the water table typically occurred at or near the surface in the spring, and up to three feet below the surface in the fall when conditions were driest. Soil textures were usually silt-loams up to three feet thick, overlying mixed cobble and gravels.

To reduce the possibility of starting a head-cut, it was important to select sections with a slope of three percent or less and with a high water table, so that the excavation would remain full during much of the year. On steeper ground, it would be necessary to bury large rock and logs across the floodplain along the upper edge of each excavated pool to prevent a head-cut from forming which would cause erosion, and fill the small wetlands with soil.

Wetlands were built in the ephemeral streams labeled as "fishless" by the British Columbia Department of Fisheries. Fishless streams generally do not contain water long enough to support fish, and may be blocked from waters containing salmon by a dam or culvert. The streams contained little, if any, surface flows during the year, an important factor potentially affecting colonization and use of wetlands by breeding amphibians.

The small mudflats were generally filled with logging slash, making them difficult to identify. Often single skunk cabbage leaves visible below piled and cut logs and branches provided the only indication of a soil-filled pool. Flagging was tied to trees and shrubs to show the edge of each planned wetland.

A small mudflat in an ephemeral stream has been filled with logging slash.

A medium-sized excavator containing a bucket and thumb attachment was used to build the wetlands. Two bucket sizes were used, a smaller 36-inch wide bucket for handling trees and stumps, and a 60-inch wide bucket for removing and spreading soil. The 200-Series Excavator was used, because of its longer reach, which allowed all operations to take place from one side of the stream, speeding construction, and allowing materials to be placed and sorted within reach of the machine.

The contractor began construction by finding a route into each worksite that would minimize crushing of planted trees and traversing of steep side slopes. The small bucket and thumb were first used to pick up and remove trees, limbs, and branches from within the flagged area. Woody material was piled out of the way, at the farthest extent the excavator could reach. Clumps of native plants were then removed and placed in piles near the woody debris. Next, topsoil was removed and placed near the flags.

Switching to the larger bucket, a depression was dug averaging 40 inches deep in the center of the flagged area. The depth of the depression was measured in relation to the elevation of the downstream edge of the mudflat. Gradual slopes were shaped around the inside edge of the dug depression. Topsoil was then spread over the bottom and sides of the excavation, and clumps of plants that were saved were replanted by the excavator. Finally, several logs and branches were set into the completed pool. Soils removed to create the depression were placed in low areas near the wetland, and spread over the surface down slope from the wetland. Some logging debris was placed over the spoil piles to help disguise them, and wheat was seeded on all exposed soils for erosion control.

Rather than build a dam across the stream channel that could erode under flood conditions, a shallow depression was created where the elevation of the inlet and outlet remained unchanged after construction. Building these wetlands was similar to excavating a sump in the bottom of a ditch where the elevation at the entrance and exit of the ditch remained unchanged.

The small wetlands can be expected to contain water long enough for amphibian reproduction, drying in late fall when the water table is lowest. They should also improve fish habitat downstream by collecting and storing spring runoff, slowly releasing waters into the ground, increasing the duration of stream flow in late summer and fall. Four or more wetlands were built in one day by using this technique.

A layer of soil is placed over the geo-textile fabric. Excess soil is blended into the hill-side, and trash hauled away.

The project is completed, ready for seeding, planting, and mulching.

This is the same site one year later.

An excavator is used to create a small depression at the outlet of a pipe draining waters from the roof at Dutch Valley Elementary School in Tennessee.

A skid-steer is used to shape the shallow depression.

A liner protected by geo-textile fabric is anchored to the bottom of the depression to form a small ephemeral wetland for amphibians.

The liner is covered with soil, the site is hand-raked, seeded, and mulched with straw. Note the flat rock placed at the outlet of the pipe to prevent erosion.

This series of small groundwater wetlands was built in the level sections of a ditch draining a large area at Temple Hill Elementary School in Kentucky.

This ephemeral wetland was built using a liner to capture runoff from the roof at St. Francis of Assisi School in Pikeville, Kentucky.

Building Wetlands of Various Types

Chapter 6

Emergent wetland along the Housatonic River in Massachusetts.

Emergent Wetlands

Many consider emergent wetlands to be the most beautiful of all wetlands. Posters, calendars, magazines, books, and television primarily feature emergent wetlands because of their appealing arrangement of water and plants. People are most likely to see wildlife using emergent wetlands, and they have become popular destinations for sightseers to visit.

Emergent wetlands may contain shallow water year around, and generally develop hydric soils and plants. Water depths may reach four feet in an emergent wetland, however an average is closer to 18 inches. They can be built in any shape, but those that resemble beaver ponds or oxbows blend with most landscapes. On more level areas of ground, emergent wetlands can be constructed to be over 40 acres in size.

When building an emergent wetland, one should consider installing a drainpipe with a water control structure, and beaver proofing the drainpipe entrance. The drainpipe allows for the efficient removal of non-native fish, adjustment of water levels for shorebird management, and access for needed repairs.

During construction, it is also wise to shape the bottom of the wetland so that all the waters can be removed through the drainpipe, much like a large bathtub. This action is critical to the successful elimination of carp from a system, as they can easily survive in small puddles of water. Shaping the wetland so that all the waters can be removed will facilitate the eradication of unwanted plants, as soils can be made to dry enough for tractors to operate in the bottom for mowing and applying herbicide.

Many emergent wetlands have been built with steep, high dams to reduce construction costs. These look artificial, and are plagued with maintenance problems. Placing a wide top on a dam with gradual 20:1 slopes will help the wetland appear natural, reduce muskrat and beaver problems, eliminate flood damage, and greatly reduce maintenance needs.

When building an emergent wetland in a sloped field, the tendency is to make the dam high enough to flood the entire field. Unfortunately, such action usually results in the creation of large areas of deep water that do not grow plants and look more like a lake than a wetland. It is better to build a number of wetlands so they are stair-stepped down a slope, much like rice-patties on a mountainside. A series of shorter dams is generally less expensive to construct than one tall dam, and the greater number of pools will increase options to manage water levels in each wetland to provide for various species' habitat needs.

Ephemeral Wetlands

Ephemeral wetlands contain water long enough for fairy shrimp, wood frogs, and tiger salamanders to breed and develop, but dry often enough so that fish do not become established. They usually fill with snowmelt and spring rains, but can also recharge after a summer storm. Ephemeral wetlands that are used successfully by a majority of amphibians for breeding generally contain water from November into August in most years.

These wood frog eggs are in a restored ephemeral wetland.

What is most important to amphibians and invertebrates that depend on ephemeral wetlands is that its waters dry on a regular basis, as drying removes fish, which prey heavily on eggs, larvae, and adults. Drying will also reduce newt, bullfrog, and green frog populations, which also require permanent water, and prey on other amphibian eggs and larvae. Sewage fungus *Saprolegnia sp.*, which can be harmful to amphibian eggs in permanent water, can also be controlled by drying.

Knowing how deep to construct an ephemeral wetland can be a challenge. It should hold water long enough for tadpoles to develop, but not so long that fish can

Restored Ephemeral Wetlands

Spring	Fall

This ephemeral wetland was restored in an old field near Beaver Creek in Menifee County, Kentucky.

This ephemeral wetland was restored in a mature forest near the Licking River in Bath County, Kentucky.

This ephemeral wetland was restored on a steep slope in a mowed field in Bath County, Kentucky.

This ephemeral wetland was established in an area where trees were harvested in Rowan County, Kentucky.

survive. Use the depth of natural ephemeral wetlands in the surrounding area as a guide when building an ephemeral wetland. It is best to design the wetland so that it will dry by the end of fall in most years, and not worry about the wetland drying in those years with above average rainfall. If one builds it shallow enough to dry in wet years, it will likely dry too early in years with normal precipitation, stranding amphibian larvae.

These tips can help one build a successful ephemeral wetland:

- Keep the watershed or drainage area small, or it may never dry.
- Do not interrupt perennial streams, or it may contain permanent water.
- Place five or less percent slopes on the dam and upper edge to help it look natural.
- Construct a spillway to handle overflow waters, so the dam does not wash out during flood conditions.
- Add a drainpipe when unsure of the optimum depth.

A drainpipe can be added to almost any constructed wetland, so that it can be managed as an ephemeral wetland. In many cases, this can be done for less money than reshaping the wetland to make it shallower.

Wet-Meadow Wetlands

A natural wet-meadow wetland contains saturated soils and hydric plants, such as bulrushes, sedges, wildflowers, and ferns, along with small pools of water where logs have blocked flow and trees have fallen. Wet-meadows generally contain gradual slopes and are found near springs, wide shallow streams, in ditches, and even at the outlet for septic drain fields.

There are many ways to build them, including:

- Modify existing ditches into a series of wide, shallow, interconnected basins with gradual slopes.
- Change the concentrated flow of water contained in a ditch into a spread-out, overland sheet flow that saturates soils.
- Build a shallow depression from soils that are high in clay that contains a packed dam and groundwater dam; avoid creating a deep pool of water by building the dam less than six inches high.
- Excavate a deep depression from permeable soils that will collect water from a parking lot or roof, but do not compact the soils. Remember to provide a spillway so floodwaters can leave the wetland.

Wet-meadow wetlands can be built to filter runoff from parking lots, roads, and roofs at schools, businesses, and homes. They can serve as above-ground, storm water management systems that mimic natural depressions and swales once present in an area.

In urban areas, wet-meadows are now being called rain-gardens, and are planted to showy species such as

This wet-meadow was restored along Beaver Creek in Menifee County, Kentucky.

Cardinal flower *Lobelia cardinalis*, wild iris *Iris versicolor*, and rose mallow *Hibiscus palustris*, along with common rush *Juncus effusus*, woolgrass *Scirpus cyperinus*, and Joe-Pye weed *Eupatorium maculatum*. Deep-rooted, warm-season grasses such as switch grass *Panicum virgatum*, eastern gamma grass *Tripsacum dactyloides*, prairie cord grass *Spartina pectinata*, and big bluestem *Andropogon gerardii* can facilitate infiltration when planted in the moist soils of a wet-meadow. These wetlands are being built as small as four feet in diameter to capture runoff at schools, and as large as one acre in old fields to provide foraging habitat for amphibians and reptiles.

Changing a ditch into a wet-meadow progression:

This deeply eroding ditch with a large watershed will be changed into an attractive, wet-meadow wetland at Lindsey Wilson College in Kentucky.

The excavator saves the topsoil and organic material.

A channel is dug down to rock and the soils are loosened in the widened floodplain. Excess soils are placed on the hillside where they will later be spread.

All rock is saved for placement in the floodplain.

The narrow ditch is widened so that flood waters can spread out and not cause erosion.

Gradual slopes are placed on the banks of the creek.

Large woody debris is broken so it will look natural.

The large woody debris is added to the floodplain.

Large rocks are imbedded in the bottom of the channel and across the floodplain.

Topsoil and organic material are spread across the floodplain.

Excess soils are blended into the hillside; they are not compacted so that trees, shrubs, and wildflowers will grow.

Wheat is sown and straw is spread for erosion control.

Straw bales are staked across the floodplain to prevent erosion until plants become established.

Forest and Shrub Wetlands

Forested wetlands can be amazing places to visit, containing large trees, mosses, and pools of water. Swamp white oak, sweet gum, and bald cypress are common in southern forested wetlands while white cedar, black spruce, and black ash are found in the north.

The most important factor to recognize when restoring a forested wetland is that trees will not germinate or grow in water. Success will only be realized by establishing trees on sites within a wetland that do not contain water and saturated soil. A visit to a natural forest wetland will show that trees are growing on rises within the wetland, not in water or on soils saturated in the summer. Investigation of swamps where bald cypress trees are growing in water will show that these trees became established under drought conditions, when the wetland was dry. One is generally not able to find a young cypress tree in these permanently flooded habitats.

There are two main strategies for building wetlands that will contain trees and shrubs. The first involves shaping the wetland and then planting the trees, the second building a dam to flood trees already established. There are some advantages and disadvantages of each technique. These are points to consider:

Flooding existing trees:

- This creates a forested wetland almost overnight.
- There must be a source of soil nearby for constructing the dam.
- The forested area may already be wetland.
- The trees that are flooded are often not those typically found in a wetland.
- Soils that are high in clay may not be available.
- The trees will eventually die once flooded.

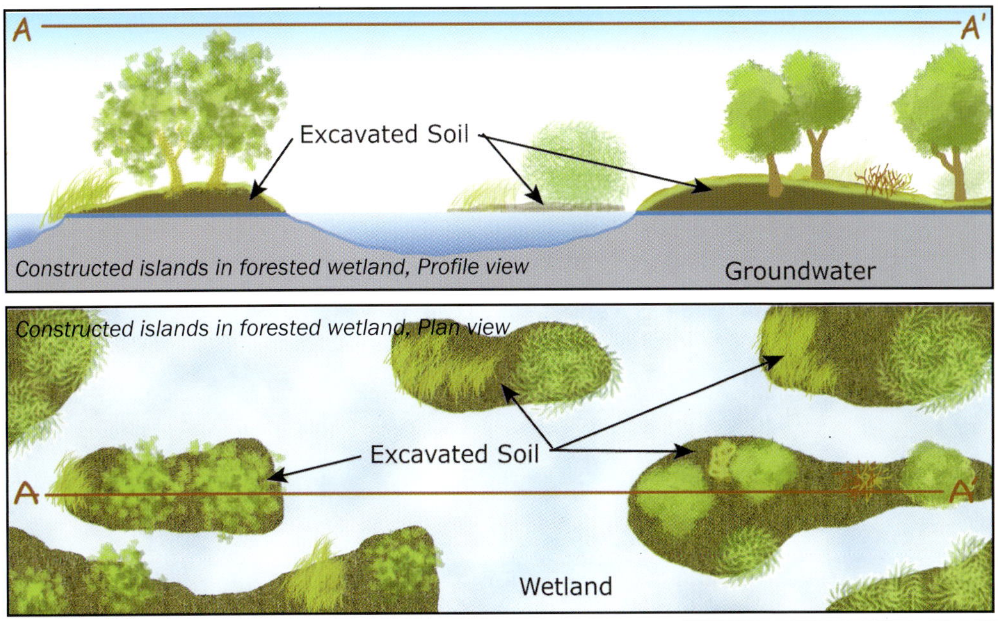

Creating a wetland before planting trees:

- This allows for the planting of trees typically found in a forested wetland.
- This increases forested acreage in open areas.
- Trees can take years to grow.
- Planted trees may not survive, because they are eaten by deer and beaver.

A successful way to establish a forested wetland involves first building the dam and groundwater dam, shaping the wetland basin, and then creating small islands within

Water-filled road ruts show vehicle impacts to this forested wetland and offer great opportunity for wetland restoration.

An excavator digs a deep trench around the worksite in search of buried drain lines for a wetland restoration project where existing trees will be seasonally flooded.

the flood pool for planting. The islands can be different shapes and sizes, and must all rise above the water level or saturated soil by one foot or more. The islets should not be compacted, as planted trees will not grow on compacted ground. Gradual slopes should be placed on the islands so they do not wash away in floods, and to reduce damage caused by burrowing muskrats. This practice of creating raised beds is the same that foresters have used for generations to grow trees in wetlands.

Another way of building a forested wetland is by constructing the dam and groundwater dam, shaping the basin, and installing a water control structure. Following construction, the bottom of the wetland is planted to trees and shrubs, and the water control structure is kept open over a period of five years or more to allow the seedlings to become established. One must be careful not to compact the soils in the wetland during construction, as trees will do poorly in compacted ground. Test for compacted soils before planting by using a shovel to dig several holes in the soil where heavy equipment was operated. If it takes considerable effort to dig a hole, the soil is compacted, and these soils should be loosened with a ripper before planting.

The water control structure can be closed to keep water in the wetland after the trees are growing, but only when they are dormant, and then for short periods of time, similar to a natural flood, to ensure their survival. The inlet of the water control structure must be modified so that beaver cannot flood the wetland before the trees are established, or during the growing season. One does not have to plant the wetland to get trees established. Over time, trees and shrubs will naturally grow in the basin, if it is kept dry.

In many fields where wetlands are established, the vegetation is dominated by tall fescue grass, which contains toxins that inhibit the growth of trees and shrubs. This fescue must be killed before planting. The combination of the herbicides Plateau and glyphosate do an excellent job. An advantage of using Plateau is that it spares native, warm season grasses and wildflowers that may be present on the site.

This five-year-old forested ephemeral wetland was created by flooding trees in an old field in Bath County, Kentucky.

Trees and or shrubs should be planted during the time of year when they can be expected to survive, not in the hot, dry weather when most wetlands are constructed. Planting one-year-old seedlings is often the cheapest and most effective way of establishing woody vegetation in a wetland, in comparison to planting larger, more costly, older trees.

Building a Wetland Where It Will Flood Growing Trees

One of the first steps in constructing a wetland that will flood standing trees is to clear vegetation from a zone that will be wide enough to build a dam with gradual slopes. A path that is approximately 200 feet wide will be needed to

This forested wetland was created by flooding trees in an old field in Menifee County, Kentucky.

A dozer is used to push the scraper, so it can better load soils. (Lindsey Moore photo)

build a two-foot high dam. Remove the topsoil from inside the zone, and stockpile it between larger trees for later spreading. An excavator and a dozer can be used to build the groundwater dam and the surface dam, making sure to dig deeply enough to cut through roots, block buried drain structures, crayfish burrows, and interrupt possible subsurface layers of gravel and sand.

There are several ways to obtain soil for the dam. One method involves clearing a zone in front of the dam and using a dozer to form the dam with the removed soil. Another technique uses the dozer to push soil into the dam from a borrow site below the dam. Or, a scraper or truck can also be used to move soils from off-site. If soil is hauled to the project, rather than depending on the site for all the soil required, fewer, if any, trees will have to be cleared, and the area of trees that can be flooded will be increased.

If a scraper is to be used to transport soil, look for an area that is high in clay near the worksite, such as in a field next to the wooded space. The source must be on a level area, as the scraper cannot climb a slope at an angle without turning over. Use a dozer to remove topsoil from the borrow area, to construct a road leading to the dam being built, and to clear a place to turn around at the other end of the dam. The scraper, then, picks up clay from the borrow area, travels to the dam, spreads the soil over the dam site, turns around and travels back over the dam, picks up another load of soil, and repeats the process.

The more efficient way to utilize a scraper is to have soil sources available on both ends of the dam, so that the scraper can travel in a figure eight pattern, and place soil on the dam with each pass. An alternative method is to have the scraper use a road parallel to the dam, so it can travel in a circle.

After the scraper has moved enough soil to build the dam, the dozers are used to shape the dam and spread stockpiled topsoil over exposed mineral soils. Wetlands may also be constructed from the areas where soil was removed to build the dam.

Using a truck to move soils for the dam can be an efficient way to build a forested wetland. A combination of two trucks loaded by one excavator can move a large quantity of soil in a day. A tri-axle dump truck can haul at least 13 cubic yards of soil, and an excavator can load a truck in ten minutes or less. Dump trucks work well when a scraper is not available; in fact, they offer certain advantages over the scraper:

- A turnaround does not have to be constructed at the end of the dam, because a truck can back up to spread its load.
- Soils can be removed from a smaller area on steeper ground.
- Soils can be hauled on improved roads and over longer distances.

Old fields that are growing up to trees offer excellent locations for restoring forested and shrub wetlands that also include an area of emergent wetland. Vegetation is removed in a zone along the lower edge of the field to obtain soils for constructing the dam, as would be done for wetlands designed to hold surface water. A dam is built along the lower edge of the field that will saturate soils and flood higher portions of the field not disturbed by construction. The area cleared for soil to build the dam will become an emergent wetland, while the higher elevations growing woody vegetation will become forested and shrub wetlands.

The wildcard that can affect the long-term success of any wetland project which includes woody vegetation are beaver. To help ensure the successful establishment of a forested wetland, modify water control structures so that they cannot be plugged by beaver. Their damage causes year-round flooding, which will kill trees, and those trees that survive may be cut for food.

Peat Land

Fens and bogs, known collectively as peat lands can be created on mined areas, log landings, and borrow pits used for road and dam construction. Bogs are basically rain-fed while fens are supplied by groundwater and surface runoff. Peat lands may develop in shallow depressions that have low permeable mineral soil or bedrock for a base, or deep organic soils that are saturated. When working to establish a peat land, select an area of land where soils can be made saturated, and where the decomposition of organic material is slow. Regions where there are less than 200 growing degree-days at cooler than 50 degrees Fahrenheit annually, or where the mean annual temperature is less than 47 degrees Fahrenheit work best.[43] Peat land can form in warmer climates where soils are saturated and water is acidic. High amounts of precipitation also appear to contribute to peat land formation.[44]

This peat land formed in a closed coal mine on the Monongahela National Forest in West Virginia. The water source is the seep at the base of the hill, marked by the small pool of water.

Here a fen formed in a reclaimed coal mine on the Monongahela National Forest at Cheat Mountain. A seep at the base of a fill slope, along with surface runoff, supplies water to the peat land.

This bog formed in a road closed 30 years ago on the Monongahela National Forest.

Sphagnum mosses form the foundation for peat lands known as bogs and fens.

Establishing a peat land on a site with some shade is fine, and can help with moisture retention and hydrologic stability. Sometimes it is easier for sphagnum to get started in a shady place, because it is wet and does not dry in the sun. There are shade-tolerant species of sphagnum, as well as those requiring full sunlight. Because sunlight is needed for a number of sphagnum mosses to survive, it is best to build the peat land large enough so that surrounding trees cannot shade it completely.

Locate a spring, seepage, or small perennial stream that is above an area containing a slope of three percent or less. One should create a wide level area of compacted soils below the water source with loose soils on the surface. The land area down slope from the water source should be shaped so that water will saturate loose soils placed over the compacted layer as it seeps downhill without forming an open stream channel.

The *ph* of the water source does not appear to be critical to the formation of a bog, although peat will develop more quickly in acidic environments. While sphagnum mosses are adapted to acidic environments, they are able to colonize saturated habitats with neutral *ph* values by growing on hummocks just above the water table, thus escaping the higher *ph* environment.[45]

Once established, sphagnum mosses release sphagnolic acid to create their own acidic microenvironment. Where water inputs to the system are very slow, sphagnum mosses may engineer an expanding acidic habitat to create a peat land.[46]

Water inputs determine whether a sedge fen or a sphagnum peat land will develop. Nutrient-rich water that moves swiftly enough to deliver nourishment to plants will result in a sedge fen. Nutritive-poor water, or water that moves very slowly through the system so that nutrients are exhausted, will lead to a sphagnum peat land. In natural settings in West Virginia, sphagnum peat lands are always surrounded by a moat, or a sort of nutrient doughnut, of richer fen or swamp. The sphagnum peat land forms in the middle, where conditions are saturated, but groundwater movement is extremely slow.[47]

Red sundew (*Drosera rotundifolia*) and light green bog club moss (*Lycopodium inundatum*) are two uncommon plants that can be found in peat lands.

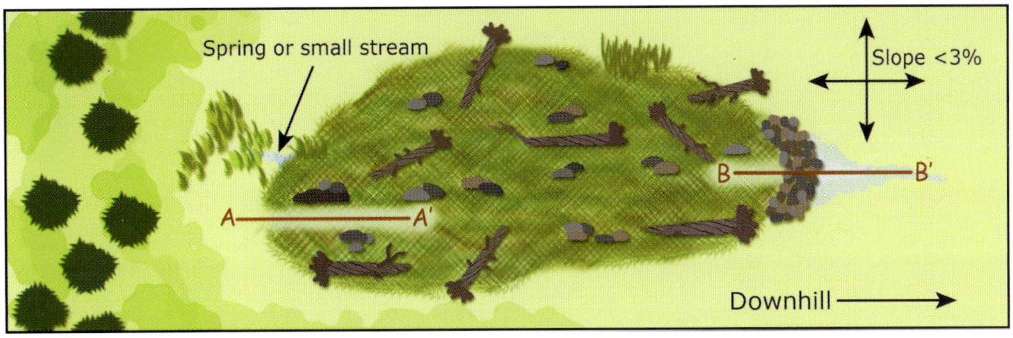

Peat land construction design, Plan view

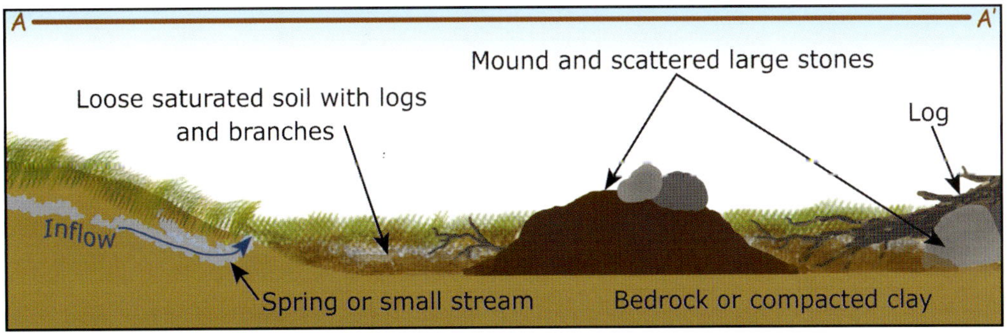

Peat land construction design, Profile view showing water inflow

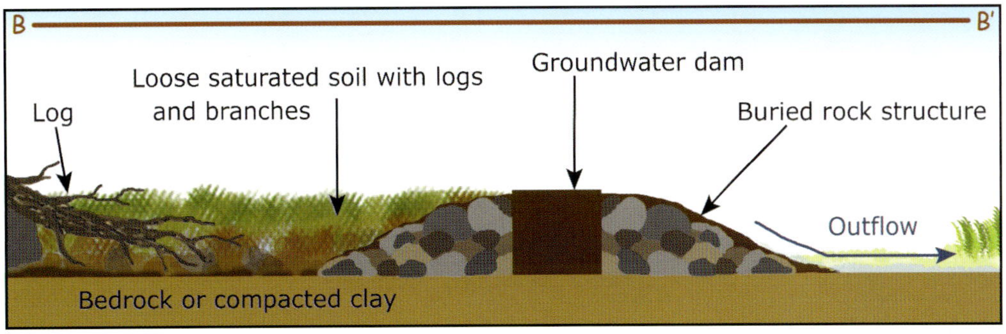

Peat land construction design, Profile view showing water outflow

The base of the bog should be rock or compacted clay that has low permeability. The area for the bog should be flattened so that both the side slopes and the downhill slope are less than 0.5 percent. When working over mining fill, it may be necessary to compact the base so that water does not soak deep into the ground. To prevent this, one may be required to build a dam and groundwater dam using the techniques previously described.

The thickness of loose soils, including topsoil and organic material, placed over the impermeable base in the new bog can vary from two to 36 inches. Soil textures can range from sand or gravel, to silt and clay. What is most important is that soils placed over the impermeable layer are not compacted.

Branches and logs can be buried in the loose soil to prevent washing, much like tree roots found in natural bogs. Logs can also be scattered on the surface to provide a platform on which mosses can grow, and to help prevent stream channels from forming, which can drain the bog.

Protect the zone where water is expected to flow from the bog by burying large rock or logs beneath the surface. These materials should be placed in a zone wide enough to prevent channels and head-cuts from forming under flood conditions.

Create a number of soil mounds on the surface of the new bog. These can be different heights and sizes from two to ten feet in diameter and from one-half to three feet high. A variety of mosses can be expected to grow on these raised areas, which should not be compacted.

There may not be a need to plant the new bog, as the spores of mosses are carried by the wind from great distances and can be expected to colonize the new bog.[48] Seeing patches of sphagnum and hair cap mosses in the area indicates that additional bog establishment may be successful. Lime and fertilizer should not be used in establishing vegetation, and the area may be planted to sphagnum mosses and other plants tolerant of low *ph* when soil moisture becomes high enough to sustain their growth.

A number of factors can damage a bog. Vehicle traffic will form ditches that can drain the bog. Water flowing in tire tracks shows that the elevation of groundwater is being lowered in the bog. Water flowing in tire tracks should be arrested by placing rocks or logs across the tracks, perpendicular to the flow of water. These structures should be buried so that water cannot flow under or around them.

Note how cattle have damaged the high dam on this constructed farm pond in Rowan County, Kentucky.

It is important to walk around the bog to identify how water is leaving it. Discovering a channel with flowing water and a head-cut can mean that the bog is threatened. Head-cuts will advance uphill into the bog, draining it over time, just as a ditch would. Action should be taken to halt the advance of head-cuts by installing buried rock structures across floodplains uphill and next to each head-cut.

A fence was constructed around three wetlands that were restored for amphibians. The ramp was designed to provide livestock with access to water in one of the wetlands.

Living with Livestock

Cows and horses can cause serious damage to restored wetlands by trampling vegetation, reducing water quality, compacting soils, flattening the bottom of the wetland, leveling the dam, changing the spillway, and shearing vertical slopes with their hooves. Constructing a fence around a wetland is the only sure way to protect it from damage. However there are measures that can be taken to reduce harm should livestock find their way into the wetland. The most important action to take is to place gradual slopes on the dam and surrounding edges during construction. This reduces vertical banks from forming should cows walk around the wetland and linger on its banks. Gradual slopes around the edge will also create a zone of saturated soil that some horses and cows are less likely to cross, forming a potential barrier to their entry. A large number of trees should also be laid in and around the wetland to reduce access.

A 12-foot wide ramp was built to provide livestock with access to water in this recently restored wetland that was also fenced. The ramp was built by spreading 12 inches of gravel over geo-textile fabric. (Sue Huddart photo)

Developing Wetlands Using Liners at Schools

Boys sample life in a wetland constructed using a liner by community volunteers at Norris Elementary School in Tennessee.

Wetlands can be constructed at schools that look and function as though they are natural. They can provide outdoor classrooms where students have an opportunity to study plant and animal life in habitats similar to those that historically occurred on the landscape.

Small, shallow wetlands with gradual slopes can be built that are safe for students to investigate. The edges of these wetlands can be approached without slipping, and the shallow depth and soft bottom are like landing on a mattress if someone does fall. A bonus with placing gradual slopes around them is the great diversity of aquatic plants that will grow in these wetlands.

Building a wetland using a rigid structure with vertical sides like that found on a plastic swimming pool is never a good plan. Amphibians and mammals cannot escape when water levels drop, and students can get hurt on the hard surfaces.

School administrators are often reluctant to authorize the installation of a wetland, which they view as a breeding ground for disease-carrying, pesky mosquitoes, in an area where so many children, teachers, staff, and parents will be exposed to them daily. Contrary to public belief, building a wetland can actually lower mosquito numbers. A wetland without fish is capable of supporting a wide variety of vertebrates and invertebrates that feed on mosquitoes and their larvae. Swallows, dragonflies, frogs, and toads will prey on adult mosquitoes during the day with bats taking over at night. The author rarely finds mosquito larvae in a wetland containing salamander larvae and dragonfly larvae.

Wetlands containing salamander larvae were found to have significantly lower densities of mosquito larvae compared to wetlands without salamander larvae.[49] The density of mosquito larvae in wetlands containing salamander larvae was 98 percent lower than in wetlands without salamander larvae. The author believes that pond-breeding salamander larvae have the potential of controlling mosquitoes and biting flies that use wetlands. Building an ephemeral wetland that does not contain fish near a wooded area is a good way of increasing the likelihood that salamanders will breed in the wetland.

A wetland project must also be accepted by residents around the school to be successful. Principals often take a proactive approach by visiting with neighbors, and by hosting an open house before building a wetland at their school. The administrators and teachers involved must be prepared for questions about safety and mosquitoes. But when informed, parents and community members can be expected to view the wetland with pride and treat it as a welcome addition to their neighborhood.

When building at schools, it is a good idea to make the wetland large enough for a class of 30 to investigate without crowding. If possible, construct a wetland whose surface area of water measures 30 feet or more in diameter. This size is large enough for students to surround without jostling each other to get near the water's edge.

As 30 of her students placed branches in the new wetland at Burnside Elementary School, teacher Anita Loveless exclaimed, "This will be great! Now I won't have to ask my husband to bring in tadpoles next year." Anita was excited about her class having its very own wetland ecosystem within sight of her school.

Finding a Place to Build

Walk around the grounds, looking for a level area, approximately 60 feet in diameter, within a five-minute

These mosquito larvae were found in a drainage ditch puddle, a more likely location for these creatures than wetlands.

Mosquitoes will breed in temporary pools that form in drainage ditches without predators such as dragonfly and salamander larvae.

walk of the school. While a wetland can be built farther away from a school, it is not as likely to be used because of the class time lost in reaching the site. Focus on areas behind the school, and look for portions of mowed and old fields that are not being used by students.

These are some places to avoid:

- Fire drill safety zones
- Next to highways
- In front of the school, as a wetland may look grown up and unkempt
- Staging and parking areas near sports fields
- Where utilities are buried
- Possible school expansion sites

Involve the school principal, groundskeeper, and other teachers in the selection of the wetland construction site. Their support is critical since they can help explain the project to parents and the school board. Explain to them that a shallow wetland with gradual slopes for safety is being designed, and that its waters are likely to dry by the end of each summer. Show them the project plan if they are interested in seeing it.

Generally speaking, the more rural the location of the school, the greater the diversity of plants and animals that can be anticipated to use the new wetland. Those built near large wooded areas in rural areas should attract a diversity of salamanders and frogs. But even urban wetlands can be expected to attract ducks, herons, frogs, and dragonflies. While wetlands built in urban areas may not attract as many species as those built in the woods, their educational value will be just as great.

Wetlands can be built at schools for different purposes. Those in woodlands that are supplied with clean water can be expected to benefit salamanders and frogs. Wetlands surrounded by un-mowed fields should attract waterfowl. Those constructed in ditches below sports fields will help remove nutrients from runoff, improving water quality and fish habitat in streams.

> *A synthetic liner must be used when building a wetland near a school, parking lot, or developed area. These places are often built up with construction fill that will not hold water.*

Testing the Soil

Assume that a synthetic liner will be needed, unless there is undeniable proof the area was not disturbed by construction of the school. Of the hundreds of wetlands the author has constructed at schools, only a handful has been successful without the use of a liner. School grounds are generally kept dry by a network of buried drain lines, and it is not wise to disable these drain lines to build a wetland.

One rarely finds undisturbed soils that contain enough clay to build a wetland near a school, so it is almost always necessary to use a synthetic liner to hold waters for maintaining aquatic soils, plants, and animals. One can expect that extensive earthmoving has been done to create the large level areas necessary for schools, parking lots, and playgrounds, and that cement, asphalt, and trees were buried in the process. Finding a little gravel, concrete, or wood in a test hole is like that tip of the iceberg which indicates much more debris is buried below.

Soil test results can be deceiving near schools and in urban areas. While they can show where soils are high in clay, they do not show what is buried deep in the ground. The unseen storm sewer, lawn drains, and rock fill can pull water from wetlands built from clay soils over the top of them.

Make sure that the project planned will not flood sidewalks, playgrounds, roads, or form low spots against buildings that will hold standing water. Use a construction level and a rod to determine where the water will go, before beginning construction, and check these elevations often while the heavy equipment contractor is building the wetland. These are important steps for all wetland work.

Synthetic Liners

A liner allows for the building of a wetland on construction fill, and on soils that are too coarse to be packed to hold water. Sandy and rocky soils, where the water table is far below the surface, are usually bypassed when searching for places to build a wetland. Using a liner is a reliable and cost-effective means of establishing a wetland on soils where no other feasible technique is available for construction.

Individuals are often tempted to use pre-formed, rigid plastic inserts to build a

This two-year-old ephemeral wetland was constructed by using a liner in Powell County, Kentucky.

This small wetland was created in 2008 in a field using a liner.

small wetland. Unfortunately, the results are generally disappointing, as there are many disadvantages to utilizing this type compared to using a flexible liner for construction:

- Amphibians, reptiles, and small mammals that fall into the water can become trapped by the steep sides and die.
- The pools are often too small for an entire class to investigate.
- They look artificial.
- They can be a safety risk, having a hard bottom and steep sides.

Determining the Size

Liners can be ordered in almost any size. Those measuring up to 40 x 40 feet can be used to construct a naturally-appearing circular wetland from 12 to 24 inches deep, with gradual sloped sides, that is large enough for 30 students to investigate without crowding. Because factories commonly seam liner materials together from eight-foot wide rolls, cost savings may be realized by ordering liners in increments of eight feet.

Marking the Construction Location

Denote where the wetland is to be built on the ground by using brightly-colored wire flags and a tape measure to mark a circle up to 40 feet in diameter. Keep the edge of the circle ten feet or more away from buildings, trees, parking lots, and utility poles, as space will be needed for equipment to operate, and for students to walk around the new wetland.

Contact maintenance personnel to check for the presence of buried utilities such as electric, gas, phone, fiber-optic, water, and storm sewers before digging. The seriousness of this step cannot be over-emphasized, so be prepared to have a teacher or volunteer parent phone the 1-800-"Dig" number in the community. Change locations if buried utilities are found on the site. For additional safety, ask that the location of all buried utilities be marked within 50 feet of the proposed wetland, and notify the equipment operator of their presence before excavation begins.

This small wetland was built a Lillooet Elementary School in British Columbia by using a liner.

Obtaining the Liner

Liners made of PVC (30 mil or thicker), and EPDM (synthetic rubber 45 mil or thicker) work well for wetland construction. The necessity of using aquatic-safe, or fish-grade liners cannot be stressed too much. The majority of synthetic liners used for roofing, landfills, and tarps are treated with fungicides and algaecides, which include chemicals that are toxic to aquatic life. Wetlands built with treated liners, therefore, will not sustain aquatic life.

A landfill operation may be a good place to contact for a possible donation of a synthetic liner. These businesses often use great quantities of liner material to create water-tight cells for garbage that cover acres of land, and may be willing to donate a piece for educational purposes. However, there may be complications with a donation, as the liners they use are very thick, difficult to transport, and are manufactured in narrow strips. These strips must be welded by a professional seaming crew, and that can be expensive. The liners may also be treated with fungicides and algaecides.

When More Soil is Needed

When building over rock, it may not be possible to excavate a depression for the wetland, or to use the rocky soil that is removed to cover the liner and make a dam with gradual slopes. This is the reason it is important to dig a test hole before heavy equipment arrives. If a 24-inch deep hole can be dug by hand without hitting rock, there should be enough soil for building the wetland. However, if rocks are encountered in the hand-dug test hole or when using a tile probe, additional soil will be needed to cover the liner. Even more soil may be needed when building on a slope.

In urban areas, it may be necessary to haul away rocky soils removed when creating the depression for the wetland. Rocky soils are difficult to blend into the surrounding landscape, and it may not be possible to mow over the rough ground.

On rocky ground, it is important to measure the slope of the ground on the desired site of the wetland. Should the slope from the upper to the lower edge of the work area exceed one percent, it may be necessary to haul additional soil that can be used to cover the liner.

Mineral soil needed to build an 18-inch-deep wetland using a 40 x 40 foot liner over rocky ground

Slope Over Work Area Percent	Soil Needed to Cover Liner Cubic yards	Soil Needed to Build Dam Cubic yards
1	30	6
2	30	13
3	30	18
4	30	25
5	30	31
6	30	37
7	30	44
8	30	50

It is best to haul any additional soil needed by dump truck. They are built in different sizes, which are known by the number of axles supporting the bed. A single-axle dump truck will haul approximately five cubic yards of soil, a double-axle or tandem nine cubic yards, and a tri-axle 13 cubic yards. Additional soil for the project does not have to be expensive topsoil; it can be mineral soil, as long as it does not contain rocks, which can puncture the liner. The soil should be dumped adjacent to the wetland construction site, and may be delivered in advance.

It is surprising how much soil is needed to cover the liner. For example, when building a wetland using a 40 x 40-foot liner on a slope that changes one percent (six-inch elevation change from top to bottom), approximately 30 cubic yards of soil will be needed to place an eight-inch layer of soil over the surface of the liner that will settle to six inches thick.

Do not be concerned about ordering a little too much soil. Leftover soil can always be used to blend the wetland dam into the surroundings. Construction and landscaping contractors are usually able to provide soil that is suitable for use on a wetland project. Soil can be used that contains a few rocks; however, sharp-edged rocks and rocks larger than a baseball must be removed by hand during construction, and this involves a great deal of effort. Topsoil is not recommended, as it often contains roots and branches that can puncture the liner, is difficult to spread, and may contain unwanted non-native, invasive plants.

Wetland with synthetic liner construction progression:

Wire flags and paint mark the perimeter of the depression to be dug. The elevation of the lowest point along the perimeter is recorded, which will be the same as the top edge of the liner. (Rachel Conkel photo)

If soils are rocky, it may be necessary to bring in rock-free mineral soil for covering the liner.

Andrea Paetow photo

The excavator digs a test hole in the center equal to (desired depth of the wetland) + (thickness of soil to be placed over the liner) in relation to the lowest point along the perimeter.

Soils are removed from the marked perimeter to the bottom of the test hole. Gradual slopes are placed along the edge with the deepest point in the center. The depression is the shape of a large satellite dish.

Barry Nichols photo

Soils removed are piled along the perimeter far enough away so they do not fall back into the depression during liner placement.

Barry Nichols photo

The size of the depression is tested before the liner is placed by using a laser level and wire flags. Flags are placed around the inside of the depression at the elevation previously recorded for the lowest point along the perimeter.

Andrea Paetow photo

A 100-foot tape is used to measure distances between flags on opposite sides around the depression. The tape is allowed to rest on the bottom. The depression must be made smaller if measurements exceed the size of the liner. The depression can be made larger if measurements are less than the size of the liner.

Barry Nichols photo

Rakes are used to smooth the depression. Soils are raked uphill so as not to fill the depression. Flags are left in place to aid in liner placement.

Mark Davis photo

Rocks are picked up and moved, tree roots are cut and the cut ends covered with soil.

Rachel Conkel photo

It takes time and hand-labor to prepare the depression for the liner.

Mark Davis photo

A layer of heavy, 8-ounce geo-textile fabric is placed in the depression. The fabric is the same size as the liner.

Mark Davis photo

The fabric is adjusted so that it covers the wire flags. The wire flags are then removed.

Mark Davis photo

A group of students are used to carry the liner. No one is allowed to walk on the fabric or liner wearing shoes.

Barry Nichols photo

The liner is placed over the fabric and adjusted so that it covers the fabric. Care is taken to remove rocks and sticks that fall onto the fabric and the liner.

John Byrd photo

A second layer of fabric is placed over the liner.

Barry Nichols photo

The fabric is adjusted to cover the liner.

A level and rod are used to check if the liner and fabric are placed correctly. The materials must reach the elevation recorded earlier for the top edge of the liner.

Barry Nichols photo

Paint spots are placed at the "top edge of the liner" elevation showing where anchors will be placed. Care must be taken not to shift the liner and fabric.

Barry Nichols photo

Materials are tacked into position by driving several spikes and washers on each side.

A 12-inch landscape spike and washer are driven at each paint spot to anchor the liner. Anchors are spaced approximately 18 inches apart.

Barry Nichols photo

The three layers are trimmed within two inches of the anchors, using sharp box cutters with disposable blades.

Lindsey Moore photo

All trimmed materials and tools are picked up and removed before covering.

The excavator begins covering the liner with soil. The soils are not compacted. Heavy equipment is not allowed on the liner or it will tear.

A small excavator can be used to place soil over the center of a large liner by rolling opposite sides of the liner inward after anchoring the center of the two other sides with landscape spikes. It is then unrolled so the excavator can travel around the outside perimeter to cover the rest of the liner.

From six to eight inches of soil are placed over the liner, including the top edge where it was trimmed.

A shovel is used to measure the thickness of soil over the liner.

Excess soils are spread along the downhill edge to avoid the appearance of a dam.

Excess soil and rock are moved off-site where space is not locally available for spreading them.

Soils are rearranged with rakes and shovels to cover thin places over the liner.

Large woody debris is placed in the new wetland.

Branches are added for perches and salamander egg attachment.

Exposed soils are covered with wheat straw to prevent erosion, increase native plant germination and survival, and reduce the possibility of introducing non-native plants.

Native plant seeds are carefully planted in and around the wetland.

Native potted plants are positioned around the shores of the new wetland. Flags can be placed at the future water level to serve as a guide for planting.

It is important to water plants until they become established.

Leaves are added to help invertebrates.

Students build a toad house from rock and clay that will be above the water level in the new wetland.

Problems with Groundwater

Water rising beneath a synthetic liner can cause the liner to float, resulting in the prompt failure of the wetland. If a high water table is detected, consider using a different construction technique, involving digging a depression that will expose the high water table, creating a groundwater wetland. However, in some locations, groundwater will rise and fall rapidly, and it will still be necessary to use a liner; otherwise, the wetland constructed would only occasionally hold water.

On sites where a widely fluctuating water table is present, it is important to install a drain under the liner before the liner is put in place during construction. The drain will prevent the groundwater from pushing the liner up from the bottom, and can be a length of slotted plastic corrugated pipe leading downhill to a ditch or stream, or a ditch filled with rock of various sizes that will carry water.

Heavy Equipment

An excavator is the preferred machine for building a wetland with a liner. The excavator should be a 100-Series or larger, weigh at least 22,500 lbs, have 80 HP or more, and a bucket at least 42 inches wide to efficiently create the depression, and then cover the liner with soil. The excavator, with its long reach, can build the wetland by not traveling over the liner, which is critical in preventing damage to the liner.

A smaller excavator can be used if the edges of the liner are rolled toward the center following placement to facilitate covering it with soil. At a minimum, the excavator should have a push blade, 54 HP or more, weigh at least 18,000 lbs, and have a bucket at least 36 inches wide.

A 4WD backhoe with a skilled operator may also be used to build a wetland with a liner, but expect the project to take longer to complete. The backhoe requires more room to operate, disturbs more ground around the wetland, and can create deep ruts with its rubber tires. A dozer alone should not be used to build a wetland with a liner. The machine disturbs a large area of ground, and cannot be used to place soils over the liner. Therefore, soils must be placed over the liner by hand, which requires a large amount of labor. A track steer alone is not recommended for the same reasons, plus the machine has difficulty digging in the compacted soils found at schools.

When building a number of wetlands with liners, efficiency is gained by using both an excavator and a dozer. The dozer is used to clear and level the site, with the excavator being used to create the depression and cover the liner with soil.

Protecting the Liner

Geo-textile fabric should be placed under the liner to protect it from sharp rocks and roots, and over the top of the liner to protect it from damage caused by deer hooves and shovels. The geo-textile layer on top protects the liner from unforeseen threats caused by someone using a tile probe or a shovel to sample soils in the wetland, which might accidentally puncture an unprotected liner. Geo-textile fabric also protects the liner from being ruined by burrowing crayfish and tree roots.

The geo-textile fabric should be ordered in the same size as the liner being used. A number of factories sell liners and geo-textile pads sewn together as a unit. In most situations, a layer of 8-ounce geo-textile fabric above and below will protect the liner from damage. One-piece geo-textile pads are easier to use and more economical to purchase than a roll of material. Thick geo-textile fabric is difficult to obtain locally in most situations, so usually has to be ordered from the factory.

One may wish to contact the local NRCS office to ask if it will donate the geo-textile fabric needed for the project. Many offices provide this fabric to landowners for erosion control and livestock operations. Some offices have narrow rolls of lighter weight fabric that, when placed in multiple layers, can be used under the liner.

In rocky soils, heavy machinery should travel over the excavated basin a number of times before placing the geo-textile pad and liner. Repeated passes will break up large rocks and help push sharp, protruding edges into the ground, reducing the possibility of puncture.

The liner and synthetic pad should be anchored along the top edge before covering with soil. The anchors will keep the liner from being pushed down into the depression into a hopeless mass while covering it with soil. Twelve-inch landscape spikes with an added washer work well for anchors. These long spikes puncture the liner and pad with ease, and can be driven into rocky soils with a three-pound sledge. Spikes should be placed around the top edge of the liner and geo-textile pads about 18 inches apart.

Covering the Liner with Soil

There are many benefits to covering a liner with a layer of soil that is at least six inches thick. The soil protects the liner from being punctured by horse, cattle, and deer hooves, and shields it from ultraviolet damage caused by the sun. Soil provides the media needed for plant growth, as well as a substrate where amphibians and reptiles can burrow to escape predators, and for hibernation. The soil covering can be used to create a gradual slope around the inside edge of the wetland for easier access. Gradual slopes also create various water depths, which in turn, increase plant diversity. Liners not covered with soil are very slippery around the edge, making it easy for students to slip and fall when investigating the site.

Locations with Cattle, Elk, or Moose

In areas where soils are rocky, or where large animals are present, additional measures must be taken to protect the liner. Heavier 16-ounce geo-textile fabric should be placed under and over the liner. This is much cheaper than purchasing a thicker liner. A layer of soil at least eight inches thick should be placed over the liner and geo-textile pads. Slopes should be kept gradual within the wetland, five or less percent, to prevent soils from sliding, or shearing caused by large animals using the wetland. Placing large woody debris in and around the wetland also appears to discourage large animal access into the wetland. It is especially important to prepare a smooth and firm bed on which to place the liner during construction, as animals with hooves stepping on a liner placed over sharp rocks could lead to failure.

Construction Steps

1. Obtain any required permits before beginning construction.

2. Check for buried utilities and avoid them.

3. Order supplies and additional soil if needed.

4. Hire an experienced contractor with a large excavator.

5. Have at least four strong individuals available to handle the liner and geo-textile pads.

6. Use wire flags and spray paint to mark the perimeter of the wetland to be built. Do this by using a 100-foot tape measure, temporarily anchoring one end to a wire flag in the center of the construction site, then stretching the tape until the radius of the wetland is reached. For example, mark the perimeter of a circle with a 20-foot radius when using a 40 x 40 foot liner.

7. Set up a construction level on high ground overlooking the worksite. The construction level

should be at least 50 feet away from the marked site. Use the level and rod to determine the lowest elevation along the perimeter of the marked circle. This elevation will be the same elevation of the top edge of the liner after trimming. The depth to be dug in the center of the circle is measured in relation to this lowest point along the perimeter of the circle.

8. Determine the depth of hole to dig in the center of the circle, for example:

   Maximum desired depth of new wetland = *1.5 feet*
   Thickness of soil to be placed over liner = *0.6 feet*
   Depth of hole to be dug in center: *2.1 feet*

9. Explain to the equipment operator that one would like a large depression dug within the area marked by the flags. The hole should be 2.1 feet deep in the center, in relation to the lowest edge, and have gradual slopes that end at the surface of the ground at the marked perimeter. The dug depression should not have a flat bottom when finished, but should resemble a large, sunken satellite dish. The top edge of the depression should be level, just like the rim of an extra-large saucer set into the ground.

10. Ask the operator to rotate the excavator and place the soils removed about six feet away from the edge of the circle, so they are in easy reach for later covering the liner, and will not fall into the hole when positioning the liner by hand.

11. After the depression has been dug, the hole must be measured to see if adjustments are needed before the liner is placed. Using the construction level and rod, record the elevation in the bottom center of the depression. Then walk up the inside slopes of the depression and place a wire flag 2.1 feet higher than the recorded elevation. Continue to mark this upper elevation using wire flags placed along the inside edge of the depression. Next, measure the distance between opposite flags by laying the tape on the bottom of the depression. The depression should be made smaller if any measurement is greater than 40 feet. It can be made larger if measurements are 40 feet or less. Leave the wire flags in place to serve as a guide when placing the liner and geo-textile pads.

12. If the soils in the depression are loose, they should be compacted by traveling over the depression several times with the excavator. This will also help crush larger rocks with sharp edges. Packing them further by having a class of kids trample the bottom also works, and makes the students feel that they are a part of the process.

13. Rarely is a depression constructed the correct size and depth on the first try, so be ready to dig deeper, to make a longer depression, or even to fill part of the hole to obtain the right size for the liner. It is generally faster to make the depression a little too small at first, and then expand it, rather than making the depression too large at the onset. Placing the liner in the depression is like burying a large satellite dish in the ground; the top rim must be level so water does not leak over a low edge.

14. Do not be concerned if measurements are somewhat less than 40 feet. As long as they are from 37 to 40 feet, the majority of the liner will be used. Hours can be spent trying to adjust the depression with heavy equipment in attempting to make it perfect, and it is just not worth the extra time. It is much easier to trim a little extra liner than to make the depression slightly larger.

15. If it appears there will not be enough soil to cover the liner, lower the profile of the wetland by excavating a depression that is deeper and somewhat wider than the original plan. By placing the liner in a deeper depression, more soil is made available to place over the liner.

16. Remove sharp rocks, sticks, or other objects from the depression by hand. Smooth the bottom of the depression by filling holes with soil. Rake sideways and uphill to flatten and level the surface, but not downhill, to avoid filling the depression. Further compact the bottom of the depression by walking and stomping over the surface.

17. Place a geo-textile pad in the depression. Make sure the fabric is next to or lies over the wire flags showing where the top edge of the liner should reach. Remove all rocks, soil, and roots that roll onto the geo-textile pad. Avoid walking on the geo-textile pad or liner until all three layers are in place and they have been covered with soil. Remove shoes if walking on them cannot be avoided.

18. Place the liner over the geo-textile pad. Remove all rocks, soil, and roots that roll onto the liner. When working with PVC, the temperature must be above freezing, or the liner may crack.

19. Place another geo-textile pad over the liner.

20. Use chalk or spray paint to mark the top edge where the liner and geo-textile pads should be anchored and trimmed. This is done by using the construction level and rod to mark an elevation that is 2.1 feet above the bottom of the depression. This operation requires three individuals. One looks through the construction level and shouts out readings, another walks around the top edge of the depression with the rod, while the third marks the geo-textile pad at the correct elevation. Start by checking elevations at the center of each side before placing marks 18 inches apart. If it is found that the materials will fall short of the desired elevation, the liner should be shifted to cover the mark. If this is not possible, the liner should be removed and heavy equipment be used to adjust soils in the depression. If the liner does not reach the desired elevation, the wetland will be

shallower and smaller than planned.

21. Prepare the landscape spikes for driving by adding a washer to each spike. Anchor the top edge of the liner by driving spikes on the marks approximately 18 inches apart.

22. The top edge of the liner should be trimmed before it is covered with soil, to minimize the amount of soil required to cover it. The wetland will also look artificial with a raised "doughnut" of soil around the edge, which will prevent runoff from entering the wetland, and create a steep slope where students can slip. Use a sharp utility knife to trim excess liner and geo-textile fabric by cutting all three layers at once. The materials should be trimmed above the anchors, leaving two inches or less between the anchor and the trimmed edge. Remove the excess liner and geo-textile fabric. Gardeners will appreciate being given these scraps for erosion and weed control.

23. Use the excavator to place from six to eight inches of the soil over the liner and pads. Do not allow heavy equipment on the liner, because it cannot support the weight, and the wetland will fail. Do not compact the soil over the liner with the excavator bucket or it may leak. Add soil so that gradual slopes are created along the inside edge of the liner. Use a shovel and tape measure to check the depth of soil over the liner while it is being covered. The soil will settle because it is not compacted.

24. It may be necessary to add another step if the excavator cannot reach far enough to place soil over the center of the liner. This involves only anchoring the very center of the liner on two opposite sides, and then temporarily rolling the unanchored sides of the liner toward the center. The excavator can then enter the depression and cover the center of the liner without traveling over the liner. Once the center of the liner is covered with soil, all ridges of soil made by the excavator in the depression should be raked smooth. The liner can at that time be stretched out to cover the depression, anchored along the top edges, trimmed, and then covered with soil by the excavator.

25. Ask the excavator operator to add more soil over thin places. Use rakes and shovels to move soils over the liner to reduce the possibility of the excavator damaging it. Use care to make gradual slopes around the inside edge, so students are less likely to slip and fall into the water. A gradual slope will support foot traffic better, and will be less susceptible to erosion.

26. Spread excess soil out from the downhill edge to help blend the wetland in with its surroundings. It is important to place a gradual slope around the outside edge of the wetland so that it can be maintained by a riding lawnmower if needed. A gentle slope of five percent works well around wetlands intended for educational use.

27. When covering the liner with sandy soils, it is best to place gradual slopes, five or less percent, around the inside edge of the wetland. Sand may slip off steeper slopes, exposing the liner to the sun and to puncture from animals.

28. Use the level to determine how waters will leave the wetland when it fills. Create a wide spillway over a gradual slope to prevent waters from flowing over the dam and loose soils that were spread. The spillway should be six inches below the top of the dam. If the spillway is on a steep slope, imbed rock in it to prevent erosion. A spillway may not be needed, if the wetland has a small watershed, and the soils used to create the dam are high in sand and gravel, as overflow waters will seep into soils above the liner.

29. Wheat provides for the rapid control of erosion, and should be seeded immediately after construction. Sow a mixture of native grasses, wheat, and wildflowers on areas of exposed soils that are above the water level for erosion control, and to reduce the potential for invasive plant colonization. Grasses such as wild rye, big-bluestem, little-bluestem, switch grass, and Indian grass work well for this purpose. Planting potted aquatic plants along the edge of the planned water line in the new wetland is also a good idea.

30. Place branches and logs in and around the wetland. Ask students for assistance in placing dragonfly perches. Search the school grounds for small branches and break them off in two-foot-long or less sections. Push them vertically into the ground within the future pool area of the wetland, making sure they do not puncture the liner. Territorial male dragonflies will use the sticks for perches when the wetland fills, and their larvae will later climb the branches to emerge. The sticks do not have to be replaced the following year, as plants growing in the waters will serve as substitutes.

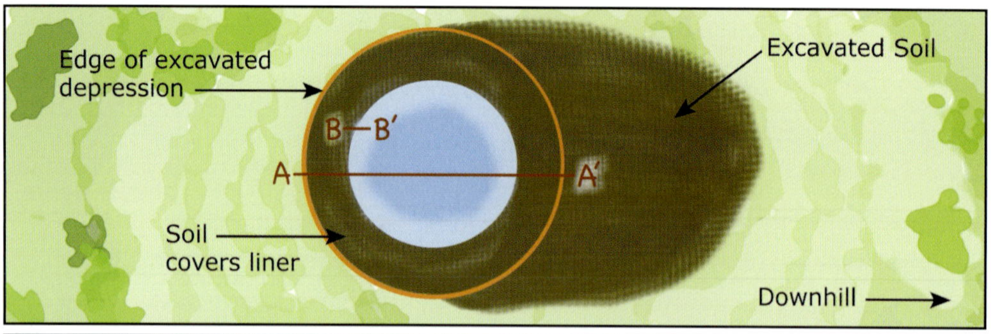

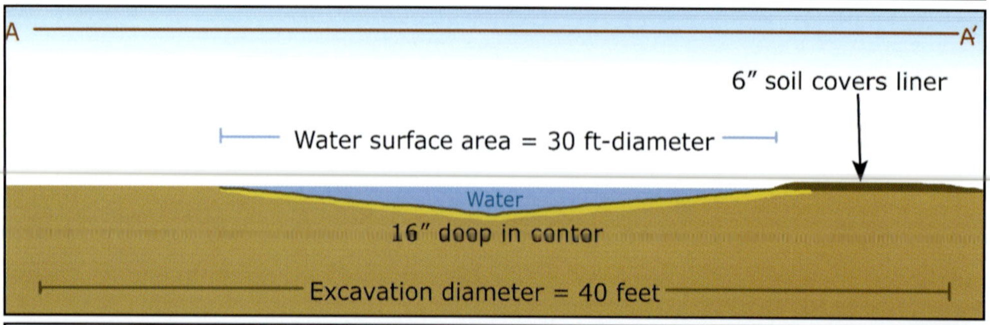

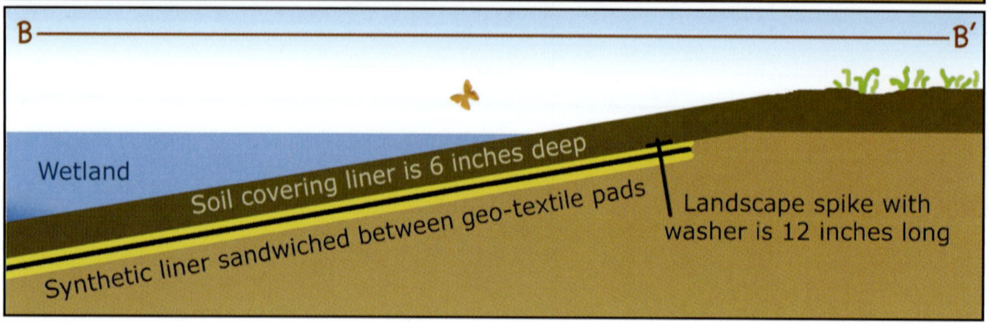

Wetland construction using liner, Plan view

Wetland construction using liner, Profile view (center)

Close-up showing how upper edge of liner is anchored and covered with soil (bottom)

Supply Needs

The supplies needed to construct a 40-foot diameter wetland with a synthetic liner are:

1. Heavy equipment: Contractor with a 100-Series Excavator, minimum 80 HP and 22,500 lbs, to move soils for creating the new wetland. Estimated time needed to complete project is eight hours. Ask the contractor if he can bring a construction level, tripod, and rod.

2. Soil: Order the necessary quantity of rock-free mineral soil by the dump truck-load when building on rocky ground.

3. Synthetic liner: PVC, 30-mil or thicker, or EPDM, 45-mil or thicker, fish-grade & aquatic-safe, one piece 40 feet wide x 40 feet long, quantity = 1

4. Geo-textile pads: 8-ounce weight or heavier, fish-grade and aquatic-safe, one piece 40 feet wide x 40 feet long, quantity = 2

   PVC liners and matching one-piece geo-textile fabric pads are available from:
   Fabseal Industrial Liners, Inc.
   42404 Moccasin Trail
   Shawnee, OK 74804
   (800) 874-0166
   http://www.fabseal.com

   EPDM liners and matching one-piece geo-textile pads are available from:
   Western Tank and Lining Ltd.
   12180 Vickers Way
   Richmond, BC V6V 1H9 Canada
   (604) 241-9487
   http://www.wtl.ca

5. Landscape spikes: twisted or smooth, 12 inches long, quantity = 85

6. Washers: one per spike (make sure they fit), quantity = 2 boxes of 50

7. Native grass seed: Plan to order one pound of native grass seed to sow on approximately 3,600 square feet of exposed soil. Possibly there is a native plant nursery serving the local area that can be found on the web.

8. Straw: Purchase 14 bales of clean wheat or oat straw to use as mulch. Plan on mulching up to 6,000 square feet of exposed soils, with one bale of straw covering an average of 400 square feet.

9. Wheat: Purchase one 50 lb bag from a farm supply store.

10. Aquatic plants: Order from one to two dozen small, native, potted aquatic plants to transplant around the edge of the waters in the new wetland.

11. Utility knives: Purchase four with replacement blades. The inexpensive, plastic, disposable kind works best.

12. Chalk or spray paint: A large piece of chalk or one can of orange spray paint

13. Shovels and rakes: One shovel will be needed for every three rakes being used by students or laborers. When working at a school with a class, it is good to have 15 rakes and four shovels available for use.

Treating Runoff

Wetlands can be used to treat runoff from parking lots and roofs at schools. This is done by directing the water from gutters and/or parking lots into the wetland under construction. There are advantages and disadvantages to this action. While the additional runoff will help maintain water in the wetland, and the wetland can be expected to clean the runoff and reduce flooding downstream, the water from roofs and parking areas may contain contaminants, and these pollutants can affect water quality, reducing animal diversity in the new wetland.

Working in Ditches

Ditches can be modified to create wetlands at schools. Some schools have so little space available that the only way possible to build is within a shallow ditch on the school grounds. A 30-foot-long section of a ditch can be deepened and widened to create a wetland. This can be done without changing the way the ditch drains. Water can be allowed to enter and leave the wetland at the same elevation as in the original ditch to avoid backing up water into buildings, onto playgrounds, or parking lots. However, it may still be necessary to use a synthetic liner, as the ditch may be located on top of porous soils, or have drain lines buried beneath it.

Blocking ditches to create wetlands requires special precautions to ensure success. People bury many things in ditches including brush, trees, drain lines, construction fill, and trash. If left beneath the new wetland, their presence often causes failure of the project.

Cues to Care

Creating a wetland for an outdoor classroom can radically change the appearance of the typically mowed and manicured school grounds. To help reduce the visual impact and gain a greater acceptance of a wetland, consider adding features to the project that landscape architects call "cues to care." A simple split rail or board fence around the outdoor classroom can show visitors, administrators, and especially the groundskeeper that the area behind the fence is something special.

More frogs and toads will be visible around a wetland if the plants growing along the water's edge are not mowed, as this is where amphibians will hide and rest. Working to keep the mowing crew from weed-eating up to the water's edge can be frustrating, but it is worth it. It is understandable to mow a path around a wetland for student access, but mowing all the plants around the wetland is not required and can prove to be harmful to the site. It is often necessary to place a number of large logs on dry ground around the wetland to prevent mowing crews from operating their machines near the wetland. Also, placing white-painted stakes next to "Please Do Not Mow" signs can help, as these stakes are the universal sign for "no mowing" to lawn crews.

Students of all ages enjoy planting soils around new wetlands.

Habitat components for a school wetland:

A. Wood duck nest box
B. Tree swallow nest box
C. Roosting box for bats
D. Cover board or metal sheet for snakes
E. Toad house made from flower pot
F. Log for turtle sunning and bird perching
G. Branches for salamander egg attachment
H. Rock pile for reptile sunning
I. Small stick for dragonfly perch
J. Sand in sun for turtle nesting
K. Oak tree for acorns, leaves, and shade
L. Island

Establishing the Preferred Vegetation

Chapter 8

Vegetation should be planted on exposed soils as soon as possible after construction to protect wetlands from being damaged by heavy rain. It is especially important to establish plants on the sloped surfaces of dams and spillways where one thunderstorm can wash away the fine soils needed for germination and growth. The best protection for exposed soils from erosion is covering them with a layer of mulch. Straw from wheat provides a highly effective and inexpensive material for this. A layer of straw protects soils from washing, and helps to retain moisture in the soil needed for seed germination and growth.

Mulching with Straw

Wetlands are generally built in the summer when it is hot and dry, often under conditions unfavorable to plant survival and growth. Seed spread on dry ground will not germinate, and if it does germinate after a summer rain, will not survive because of continued hot and dry weather baking the ground. A layer of straw makes it possible for seed to develop and grow in the summer, even under drought conditions. Wheat or oat straw is preferred over hay for mulch. Straw contains few, if any, weeds, and is much easier to spread than hay.

There are different zones of vegetation to establish in a wetland project, with each zone separated by its expected soil moisture level. First priority should be placed on establishing vegetation on the driest zones above the planned water level in the new wetland. Dry zones on slopes are most susceptible to erosion, requiring immediate action to protect them from damage caused by thunderstorms. The lower and more level zones that will be saturated when the wetland fills are less likely to erode, providing more time for growing the preferred plants.

Wheat should be seeded on exposed soils in all soil moisture zones before they are covered with straw. Wheat can be expected to germinate within three days after rain, with its roots holding in place soils even under flood conditions. Seeding to wheat should be done even when no rain is predicted. Heavy dew often provides enough moisture for wheat to germinate and grow. Oats can be used if wheat is not available. These species are short-lived, non-invasive, and provide rapid, inexpensive, temporary erosion control.

Annual or perennial rye is not recommended due to allelopathic properties that can affect the establishment of native plants.[50] Annual rye can also persist for three to four years, competing with preferred species that are planted.

To protect wetlands from erosion, seed and mulch each one as it is completed rather than waiting until the end of the project. Small wetlands should be seeded and mulched with straw the same day they are built. Larger wetlands should be seeded and mulched in stages as sections of the dam are completed.

Straw can be spread by hand or with a blower. For small projects, it is generally easier and cheaper to spread by hand. While only one person is needed to spread straw by hand, three to four individuals are required to use a blower. One worker drives the truck or tractor, another opens the bales and feeds straw into the blower, while the third operates the nozzle on the blower.

Straw may be baled with string or wire. String-tied is the most common. Wire-tied bales are usually heavier, longer, and contain more straw than string-tied bales. A string-tied bale will weigh approximately 20 pounds, while a wire-tied bale will be about 30 pounds. One bale of string-tied straw will mulch approximately 400 square feet, a wire-tied bale, 530 square feet. Straw should be spread to cover 90 percent or more of exposed soils. It is of prime importance that sloped areas be mulched, because they are the most likely to erode.

Wheat should be sown on exposed soils the same day the wetland is completed to help control erosion and colonization by non-native plants.

Wheat straw should be used to mulch exposed soils on completed wetlands before the next rain to control erosion. (Sarma Liepins photo)

A dedicated worker can be expected to hand-spread about five bales of straw per hour, or about 40 bales in a day. A team of three individuals operating a straw blower can spread 15 bales an hour, or about 125 bales in a day. In order to use a blower, it must be possible to drive a truck or tractor around the worksite. Soils must be firm enough to avoid miring vehicles, and there must be room to maneuver pickups and trailers. While straw blowers are available for rent in most communities, it is important to be sure the project is large enough to warrant their use. These machines are prone to having mechanical problems, so it is best to plan extra time for repairs while trying to finish a job.

Care should be taken to avoid damaging the new wetland while moving straw around the worksite. Soils may become compacted if workers are allowed to drive vehicles with abandon to deliver and spread straw. Compaction can be avoided by hand-carrying straw, or by using OHVs to move supplies around the worksite.

Consider in advance how to purchase and transport straw to the worksite. Small quantities of straw can be bought from a farm supply store and hauled to the worksite in a pickup truck. A pickup with a full-size bed can safely haul about 40 bales of straw. For larger projects, it is generally less expensive to order from the supplier. Straw contractors will often deliver, unload, and stack a large truckload containing from 300 to 400 bales at no additional charge.

Planting Native Species

Tall fescue grass seed is commonly used to stabilize drier zones subject to erosion because it is highly effective, inexpensive, and readily available. However, its use on wetland projects is discouraged because of conflicts it causes in establishing native plants, and providing for wildlife use. Fescue contains chemicals that inhibit the growth of native wildflowers, shrubs, and trees. Areas sown to the Kentucky 31 variety of tall fescue along with the endophyte-free varieties of fescue will form dense, persistent, monotypic stands of grass that are barriers to other species' survival and growth. Conversely, areas sown to native grasses and legumes are more open, diverse, and eventually become colonized by trees and shrubs.

Placing a layer of topsoil over sloped surfaces and zones that will remain above the water level will speed the vegetation process. Topsoil will contain seeds that will grow along with planted species, and the roots found in topsoil act like mulch to reduce washing.

Native plants, such as switch grass, Indian grass, big bluestem, wild rye, partridge pea, and Illinois bundle flower, can be sown on drier zones at the same time as wheat. When planted in early summer, they will germinate and grow, providing lasting protection for the soil. Prairie grasses can be difficult to sow by hand, because they are light and fluffy, presenting a problem on windy days. Commercially available hand-held warm season grass seeders can be used with success to sow these seeds. The de-bearded seed is now available and can be spread by using a conventional, hand-held seeder.

A diversity of aquatic plants naturally colonized this emergent wetland constructed in 2003.

Visit the Natural Resources Conservation Service Plants Database
http://plants.usda.gov/index.html
for information about native plants, invasive plants, and noxious weeds in your area.

Bright sedges add beauty to this restored emergent wetland.

Marsh marigold adds color to this forested wetland near Elkins, West Virginia.

Planting zones of a wetland project that will become saturated or covered with water is not required for producing a successful wetland, but it can help. Many aquatic plants can be expected to grow naturally in a constructed wetland over time. April Haight found that 50 species of obligate wetland plants grew naturally in restored wetlands near Beaver Creek on the Daniel Boone National Forest in eastern Kentucky within five years of construction.[51] A diversity of aquatic plants has also been found to grow naturally in isolated wetlands constructed on mountain ridge-tops in eastern Kentucky.

Planting saturated soils and future inundated zones in a new wetland can reduce possible colonization by cattails and non-native plants. After a wetland is built, it is generally the aquatic plants in the surrounding area that have their seeds transported onto the exposed soils first that will dominate the site. Too often, this tends to be cattails because of their prolific production of light, wind-blown seeds. Sowing native bulrush and sedges following construction can help more desirable species to succeed.

Common aquatic plants, such as lamp rush *Juncus effusus* and wool grass *Scirpus cyperinus* can be planted by spreading their seeds over exposed soils. Their spiklets can be collected in late summer when seeds are ripe by simply breaking them off, and then spreading them over the planting site. For germination to occur, seed must be in contact with moist soil, receive direct sunlight, and overwinter on the soil surface. As long as moist conditions can be sustained and early competition reduced, seedlings will develop the following spring.[52]

Another reason for planting is to more quickly establish attractive plants that would normally take many years to grow in a wetland. Planting can add color and diversity.

Adding potted blue flag iris, cardinal flower, monkey flower, rose mallow, and buttonbush is an excellent way to enhance the beauty and contrast of a newly-completed wetland.

Timing is everything when it comes to establishing aquatic plants in a wetland. It is best to plant in spring when the soils are damp, and laborers can see where water is standing. Planting in late summer when it is dry is a guarantee of failure.

Obtaining aquatic plants from donor populations can be a low-cost, efficient way of improving diversity in a new wetland. Private landowners generally give permission for individuals to transplant aquatic plants growing in roadside and drainage ditches. Ditches are excellent places to obtain plants. But, because they are regularly cleaned with heavy equipment and often sprayed with herbicide, they should be considered high-risk habitats. The margins of constructed farm and ranch ponds are also good sources for aquatic plants.

Locating sources for aquatic plants is seldom a problem with the many commercial nurseries that specialize in wetland mitigation projects available. Many states have native plant nurseries that offer an assortment of plants at reasonable prices. For best results, the aquatics planted should be native species that are grown in the same region the project is located. Potted plants can also be expected to survive and grow rapidly on wetland projects. A post hole digger is a useful tool to create holes for setting potted plants. Planted aquatics may suffer high losses from feeding by Canada geese, muskrats, deer, and beaver, so it may be necessary to protect plants with shelters and temporary fences.

Establishing Trees

Trees require soils that are not saturated for germination and growth. In wetlands, a majority of trees will be growing

Trees can be expected to grow naturally on the mounds of soil placed in this newly constructed wetland.

Buttonbush is an excellent choice for planting in a restored wetland, its beautiful flowers attracting a variety of butterflies and moths.

on higher ground, not in standing water. Even cypress must germinate and start growing before a wetland is filled. Trees grow best on drained areas, which are generally found on raised ridges and mounds in wetlands, often adjacent to standing water and saturated soils.

While bottomland hardwood trees, such as swamp white oak and pin oak, can survive in soils that are saturated for part of the year, they cannot be expected to take root in wet ground. Therefore, it is important to change locations to a drained site if water is observed seeping into the hole being dug to plant a tree on a wetland project.

Many trees can be anticipated to seed on their own, germinating on exposed soils following wetland construction. Trees with light seeds transported by the wind, such as red maple, box elder, sycamore, spicebush, Virginia pine, and green ash can be expected to colonize a site without planting. On many locations, costs can be lowered by waiting for trees to grow spontaneously.

Trees are usually available from a nursery as bare-root seedlings or containerized stock. Bare root trees are inexpensive to purchase, and are easy to plant by using a metal planting bar or a planting machine pulled by a tractor. Bare-root trees are generally one year old, and are sold in bundles packaged in large paper bags. The seedlings must be kept cool and moist until they are planted. The best way to accomplish this is to keep them in a refrigerator until the day of planting, and then to transport them to the worksite in coolers filled with ice.

Containerized trees are sold in pots or wrapped in burlap. One-year-old trees are grown in small pots, whereas saplings are available in much larger pots. The small containerized trees can be planted by hand with a shovel, while a motorized auger is needed to plant larger trees.

Bare-root and containerized trees have their advantages and disadvantages. While bare-root trees are inexpensive and easy to plant, their survival rate may be lower. While the survival rate of containerized trees is generally high, height growth after five years is often less than that of bare-root trees.

Steps to planting bare-root trees:

A drained area is selected for planting a one-year-old, bare-root pin oak seedling.

Grasses that may compete with the seedling are removed by using a fire rake to scrape an area over two feet in diameter.

A shovel with a long and narrow blade is used to create the planting hole.

The blade is pushed deep into the ground.

The handle is moved back and forth to make the hole wide enough for tree roots.

Seedling roots can now be eased into the hole.

The hole must be deep enough so that the roots will extend fully downward and not be crushed into a J-shape.

The roots are guided to the bottom of the hole and then raised so that the tree will be planted at the same depth as where it was grown in the nursery.

Soil is deposited into the hole around the roots.

The tree is kept in place by heeling in soil around it.

both rhizome and roots are removed. Pulling quickly simply breaks off the leaves, leaving roots and rhizomes in place that will sprout and grow. Removing cattails by hand is difficult and time-consuming, and therefore, practical only on small projects.

- Shade the wetland: Cattails can be reduced, and eventually eliminated, by shading a wetland. Some small wetlands built in timber harvest areas on the Daniel Boone National Forest in Kentucky are colonized by a dense growth of cattails. The cattails will disappear in about 20 years when surrounding tree seedlings grow up around the wetlands, shading their waters.

Dense growth of cattails was a problem in this five-year-old emergent wetland prior to treatment in May, 2006 with a formulation of glyphosate labeled for aquatic application. Also note the ephemeral wetland in the background covered by cattails.

This photo taken in June, 2006 shows how cattails were controlled by applying glyphosate herbicide with a hand-held garden sprayer. Applying the herbicide took 20 minutes at a cost of $30. The gray tub on a post in.the wetland is a Canada goose nesting structure.

A diversity of sedges and rushes has replaced cattails in the treated wetland by July, 2009. Note a similar reduction in cattails in the background wetland, which was mowed when dry in the fall of 2007, using a tractor and bush-hog.

Designing Wetlands for Wildlife

Steps can be taken to greatly improve the odds of certain wildlife species using a new wetland. Often even a minor change in habitat can make a big difference in attracting animals to a new wetland. Just as it is not possible for a natural wetland to provide habitat for all wildlife species in an area, neither should a constructed wetland be expected to do the same. Designing a wetland to provide habitat for one species or a group of species will increase the chances of success.

Fallen Trees

Visitors to a forested wetland soon notice where trees have fallen into the water. Watching a log in a wetland, they will likely see turtles sunning, birds perching, and muskrats feeding from the safety of these little islands. Consider adding fallen trees to wetland projects. They are often available after preparing a site for wetland construction. They can be dragged by pulling them with a dozer and chain. An excavator with a thumb attachment can pick up a fallen tree and place it almost anywhere.

When clearing trees, leave the roots attached rather than removing them with a chainsaw. Root masses and the soils surrounding them will grow a diversity of plants and provide habitat for turtles, frogs, salamanders, and dragonflies. Fallen trees should be anchored with soil so they do not wash away. This can be done by excavating a hole large enough for the lower half of the root mass and then grading soils around the roots and over a portion of the trunk. Trees will stay in place and look as if they naturally fell into the wetland.

Large fallen trees can also be used to make wetlands and the access to wetlands less attractive to the users of OHVs. These can turn wetlands into muddy, compacted racetracks devoid of life in a single weekend. Placing large diameter trees in and around wetlands, and across access trails can help protect these sites from damage.

Standing Trees

Leaving live or dead trees standing in constructed wetlands can pay big dividends for wildlife. Birds use trees in wetlands for perches and for nesting. Herons build rookeries in patches of trees flooded by wetlands.

For added diversity, consider replanting trees upright that were cleared when building the wetland. These replanted trees will die and become snags. To plant these snags, ask the equipment operator to dig a hole where the snag is to be located. Make the hole large enough to hold the entire root mass. The root mass should be buried several feet deeper than it was when the tree was growing to make it is less likely to topple. Pack soils solidly around the tree. It takes an average of 30 minutes of heavy equipment time to plant a fallen tree for a snag, and is well worth the effort and the little added expense.

Toads

Toads are generally most successful breeding in shallow, open water wetlands that are free of fish. Small wetlands located in sunlight appear to be favored sites for laying eggs.

A toad house is being constructed near the water's edge in a newly-built ephemeral wetland.

Large rocks unearthed during construction can be placed around the new wetland to provide places for students to rest, and toads to hide.

Toads can often be found buried in soils within and surrounding wetlands. However, they are unable to burrow if soils have been compacted by heavy equipment, motor vehicles, or livestock. While the Eastern spadefoot toad may breed in the small pools found in residential areas, it is unable to burrow in mowed lawns, reducing habitat available for the species.[53] Soils that are compacted within and surrounding wetlands should be loosened to benefit toads. Loose soils can be spread in and around the wetland basin during construction to provide habitat for toads.

American and Fowler's toads can be encouraged to seek refuge near wetlands by placing stumps, pieces of bark, logs, rock piles, and constructed toad houses near the water's edge. Flat stones, bricks, two by four-inch wood pieces, or even a partially buried flower pot can provide the damp, dark shelters they use.

Boreal Toad

The boreal toad can be found breeding in wetlands that contain shallow, warm water and are protected from wind. The wetlands are generally in sunlit locations with waters only six inches deep. These wetlands contain areas that are free of vegetation which the toads use for basking, while zones of aquatic plants provide larvae with cover. High winds can affect breeding success by breaking up egg masses.

Special landscape conditions are necessary for building wetlands that can be used by the boreal toad for breeding in Colorado on the Front Range. The most important is the presence of a spring. Runoff from rainfall is not enough to maintain waters in shallow water wetlands over the summer months. Evaporation can dry six-inch-deep standing waters in a matter of days. Therefore, the constant input of groundwater from a spring is critical for maintaining shallow waters over a long period.

Many of the springs across the Front Range have been altered for mining and ranching activities. Wells have been dug at the emergence of springs, with overflow waters directed underground in pipes. In addition, natural wetlands supplied by spring waters have been eliminated by ditching. Springs generally form streams, and in many cases, these streams, which were naturally sinuous, have been changed into straightened ditches, offering opportunities to restore small wetlands that can help the boreal toad.

The western toad, also known as the boreal toad, will successfully breed in constructed wetlands.

This wetland was created from a spring on a mined area within the Roosevelt National Forest in Colorado. Boreal toad larvae are visible between the two large rocks.

These green frog tadpoles are preying on Jefferson and spotted salamander eggs. Hatched the previous spring, the large tadpoles overwintered in the permanent waters of the wetland. (Chris Yearick photo)

Wood Frog

Wood frogs appear to breed most successfully in wetlands that are shaded by trees, and which dry every fall. The shading reduces cattail abundance, which can crowd open water needed for egg laying. The annual drying is important for two reasons; the first is that it prevents fish from surviving, which eat adult wood frogs, eggs, and tadpoles. The second is that green frog tadpoles, which take two years to develop, are prevented from surviving the winter in the wetland. These tadpoles hatched the year before can prey heavily on wood frog eggs laid in early spring.

Wetlands built for wood frogs can be quite small and shallow. In the eastern United States, wetlands only eight feet across and 12 inches deep are successfully used for breeding. An important consideration is that the wetland must not dry before the wood frog tadpoles fully develop. To determine the depth to construct a wetland for wood frogs, it helps to measure the depth of wetlands in the surrounding area in which they successfully metamorphose.

To reduce the potential for cattail growth in new wetlands built for wood frogs, select sites in wooded areas where mainly under-story trees are removed for construction. Should this not be possible, and the new wetland will be exposed to sunlight after construction, planting trees around the water's edge can provide future shading.

This red-legged frog is using a restored wetland on Vancouver Island, British Columbia.

Red-legged Frog

The red-legged frog is unable to compete with introduced populations of bullfrogs. Therefore, management for the species must discriminate against the bullfrog while providing conditions suitable for red-legged frog breeding. While bullfrog tadpoles require two or more years to develop, the red legged frog can metamorphose in one year. However, because the red-legged frog takes all spring and summer to develop, ephemeral wetlands should be constructed to contain water into September before drying.[54]

Sunlight that warms waters is important to red-legged frog larvae development, explaining why they often breed in wetlands located in timber harvest areas, near the edge of fields, and in transmission line ROWs. Red-legged frogs often lay their eggs along the northern edge of a wetland, where they will receive the most afternoon sun in shaded areas.[55]

Four-toed Salamander

A woodland species, female four-toed salamanders tunnel into the clumps of moss growing at the base of trees overhanging the water's edge, on vertical banks, and logs in natural and constructed forested wetlands, which do not contain fish, to lay their eggs. Guarding their eggs until they hatch, larvae drop from the mosses into water to develop. Where soils are high in clay, white tailed deer can create the vertical banks used by the four-toed salamander around wetlands by shearing soils at the water's edge when drinking.

This turtle nesting beach was constructed of sand with a pea gravel base at the Toronto Zoo near an old farm pond used by the Blandings turtle. (Julia Phillips photo)

Breeding habitat can be increased by creating vertical surfaces along the edge and within zones of open water in wetlands. Vertical banks can be created that are six inches or higher. Logs placed in waters will also grow mosses, as will the bases of trees planted at the water's edge.

Mosses overhanging waters in an ephemeral wetland may be used by four-toed salamanders for breeding.

Carefully peeling back mosses can expose eggs laid by the four-toed salamander.

Four-toed salamanders generally will not breed in constructed wetlands until mosses are growing, which can be 15 years or more following construction. Mosses can be planted along the water's edge in constructed wetlands to speed use by four-toed salamanders. The common fern moss *Thuidium delicatulum* will quickly colonize moist soils on exposed banks when inoculated with moss plant fragments. Spreading a slurry of moss fragments and buttermilk on moist surfaces can promote their growth.[56]

Marbled Salamander

The marbled salamander lays its eggs beneath logs in wetlands that are dry in the fall. This strategy allows the eggs to hatch soon after waters fill the wetland, explaining why their larvae are generally larger than other salamanders that breed in the spring. Adding logs, six inches or greater in diameter to natural or constructed ephemeral wetlands in woodland areas can improve breeding habitat for the marbled salamander.

Spotted Salamander

The spotted salamander will breed in ephemeral wetlands located in woodland areas, and in small openings within the forest. The spotted salamander, along with other species, such as the Jefferson salamander, attaches its eggs to branches that have fallen into the water. The eggs have a greater chance of hatching when suspended in the water, as oxygen levels can be lower in the bottom sediments of the wetland. Adding branches to wetlands is a good way to help many pond-breeding salamanders.

In order to create a turtle nesting area the excavator digs a hole above the water level in the new wetland. Sand that was found within the pool area of the new wetland, as marked by the orange flag in the foreground, will be used to fill the hole.

Sand that was discovered while restoring a wetland is moved by using an excavator.

The sand is placed in a hole that was dug above the water level in the new wetland.

Sand is added until the hole is filled. Exposure to sunlight is necessary for the turtle eggs to hatch.

Turtles

Turtles lay their eggs in loose soils that are exposed to direct sunlight near wetlands, lakes, and rivers, where their eggs can remain in the ground a year or more before hatching. It is important that nesting habitat receive sunlight for most of the day, so temperatures remain high enough for egg development and hatching. Nesting habitat can be maintained by removing plants shading these areas. Fields near wetlands can be improved for turtle nesting by disking them, and then not planting. Piles of sandy-loam soils can also be placed near wetlands to provide nesting sites.

Logs and large rocks can be added to wetlands to provide turtles with basking and resting sites. In flood-prone areas, logs should be anchored with soil to keep them from washing downstream.

Add or create loose soil in the bottom of wetlands to provide habitat for turtles overwintering. Turtles bury into wetland soils to evade detection by predators, such as the otter. Young turtles can be found buried in the saturated soils of wet-meadow wetlands when it is hot and dry. Some will continue to stay buried even if the wetland dries.

Springs that are sometimes uncovered when restoring wetlands offer great opportunities to provide habitat for hibernating turtles. A depression lined with loose soils should be created over the actual spring to improve turtle habitat. It is important not to compact soils in the depression so turtles can burrow, and waters from the spring will continue to flow.

Wood Ducks

Wood ducks will use small wetlands for feeding, especially if they are located beneath the overhanging branches of oak trees producing acorns. Since white oaks often grow on soils that are high in clay, it is sometimes possible to construct wetlands so that acorns from adjacent trees will fall into water, benefiting wood ducks. Even wetlands under 30 feet in diameter can be used by large numbers of wood ducks when overhanging oaks are present.

Wood ducks require tree cavities in which to lay their eggs. These natural bird boxes are generally located within one-half mile of wetlands containing vegetation dense enough for ducklings to hide when pursued by predators. Placing nest boxes specifically designed for wood ducks near wetlands is an effective way of increasing their numbers. These boxes should be placed in wooded areas near the wetlands, so that no box is visible to any other. Wood ducks are adept at finding boxes in the woods, so they do not have to be placed in open areas at the water's edge.

Hiding wood duck boxes reduces the loss of eggs from predators, such as raccoons and snakes, which can learn to identify these boxes as a source of food. Reducing the visibility of boxes will also reduce "dump nesting," a behavior that involves more than one wood duck laying eggs in the same nest box, usually a highly visible box, which results in none of the eggs hatching. These boxes are also used by the hooded merganser and golden eye duck for nesting.

Wood duck nest boxes made from plastic have advantages over those made from wood. Because they are lighter in weight, they are much easier to place than wood. Placing a wooden box in a tree can be dangerous, requiring the use of a rope and pulley to hold the box in place while it is being nailed to the tree. Plastic boxes will not rot, so they last much longer than wood. The backs of wooden boxes will split as the tree grows, destroying the box. While plastic boxes will also eventually fall from trees as they grow, they may be easily replaced without damage.

Canada Geese

A Canada goose normally lays her eggs in tall vegetation along the edge of a wetland, where the nest can be disturbed, and eggs eaten by raccoon, fox, skunks, or dogs. Islands in wetlands provide a safer location for geese to nest, and building them in emergent wetlands can help increase goose production. A low-cost and highly effective way of raising the rate of goose nesting success involves placing elevated plastic tubs in and near emergent wetlands. A nesting tub mounted on a post one foot or more above the water or four feet above the ground will be inaccessible to most predators.

Managing open grassy fields near wetlands will provide geese and their goslings with areas to feed. Grasses kept short by mowing and green by fertilizer application are favored sites for goose foraging, especially in areas where hunting is restricted and dogs are restrained.

Shorebirds

Shorebirds are remarkable in their ability to find mudflats during spring and fall migration, even when flying at high elevations. Wetlands built with gradual slopes can provide important resting and feeding habitat. The slopes in the bottom of wetlands built for shorebirds should be gradual, averaging one or less percent. Wetlands built on steeper ground can be improved for shorebirds by creating level benches that are stair-stepped down slopes into deeper waters.

Recently exposed moist mudflats receive the greatest use by feeding shorebirds. Mudflats dry quickly, and will grow plants before long, receiving little, if any, use by shorebirds. Water control structures placed in wetlands allow for the gradual dropping of water levels over a period of weeks and months to expose new mudflats. In years with abnormally high precipitation, having the ability to change water levels in wetlands will maintain mudflats for more days than would occur naturally through evaporation.

American Woodcock

The American woodcock is known for its courtship antics involving a spiraling aerial display and the strange "peenting" sound it sings to attract a female. Its extended beak makes it a specialist in catching earthworms found in old fields and alder thickets. The woodcock depends on areas of alder, found in scrub-shrub wetlands for feeding and nesting. Alder is often found in old fields originally created by draining wetlands. It begins growing when fields are no longer mowed, plowed, or grazed, and when drainage features are not maintained. Alder can dominate wet fields for years, but will eventually be replaced by trees, losing its value to woodcock. Alder can be maintained on a site by periodically cutting trees and shrubs, and allowing them to sprout. A variety of techniques is used to regenerate alder, including hand-cutting, mowing by tractor and/or bush-hog, and shearing by dozer when the ground is frozen.

Bats

Bats make extensive use of wetlands for drinking and feeding. Small wetlands constructed on dry mountain ridges are used by eight species of bats on the Daniel

This small wetland was created in a road on a ridge-top to help close the road to vehicle traffic. Located within one-quarter mile of caves used by federally endangered Indiana bat and Virginia big-eared bat, it receives use by both species.

Boone National Forest, including the federally endangered Indiana bat *Myotis sodalis* and the Virginia big-eared bat *Plecotus townsendii virginianus*. Pregnant and nursing bats frequent these water sources during summer months when their need for water is greatest.

Building wetlands on dry mountain ridge-tops along roads and logging skid trails can create conditions where large numbers of bats can be expected to use them. Road corridors are used by the bats as flight paths, and these open routes provide unobstructed access to the water in wetlands for drinking. Building wetlands in roads provides an effective way of closing them to vehicle traffic, and improving habitat for bats.

When constructing a wetland for the primary purpose of watering bats, it is best not to remove larger trees that are shading the site. Maintaining shade over the new wetland will reduce the possibility of a cattail invasion, because dense cattail growth can block bat access to waters for drinking. When the wetland is built away from a road, a flight path should be created for bat access. Where there is concern that waters are available to bats throughout drought, the wetland should be made deeper, and these deeper waters may be less likely to support cattails.

Dragonflies

Dragonflies are one of the most menacing predators of mosquitoes. Dragonflies that breed in the still waters of wetlands are primarily territorial, where males can be observed perched on plants over waters. Females will leave upland areas when they are ready to lay eggs, finding males for fertilization at the wetlands. The female lays her eggs in the water, often attaching them to grasses just below the surface of the water.

Underwater muskrat burrow entrances are visible in a created island within a constructed wetland being drained on the Hoosier National Forest in Indiana.

Dragonflies make the greatest use of wetlands that are in sunlit locations. They will, however, also use wetlands in forests that are not completely shaded. According to Dr. Austin Jenkins, one of the dragonfly's greatest needs is upright vegetation within the wetland that can be used by larvae to scale when surfacing. He says that "bulrushes, sedges, cattails and shrubs surrounded by water provide the safest sites for larvae to emerge." Placing small sticks upright in a newly-created wetland will benefit dragonflies until aquatic plants become established.[57]

Some of the rarest of dragonflies are specialists that breed only in ephemeral wetlands without fish. Establishing wetlands that periodically dry can offer visitors colorful displays of these flying acrobats seldom witnessed by others.

This emergent wetland in a sunlit location receives much use by dragonflies. It was created on mined land with a high water table for $1,500.

Muskrats

Islands of soil can be placed in wetlands to provide sites for muskrat dens. The islands should be constructed from loose, unpacked soils, with steep sides, and extend above the water level by two feet or more. Muskrats will tunnel into these islands from underwater to create dens for resting and breeding.

Crayfish

There are many reasons to manage wetlands for crayfish. They help form an important part of the food web, with over 240 animal species known to use them for nourishment.[58] Crayfish can be very abundant in an area; one study in eastern

Texas found a concentration of 63,504 crayfish mounds per hectare in a wet field.[59] Crayfish will not damage wetlands, providing the techniques described in previous chapters for wetland construction are followed.

The burrows made by crayfish provide habitat for other species. On two occasions, Wildlife Biologist John MacGregor observed where spotted salamanders laid eggs in the waters of a crayfish burrow. He has found black rat, corn, ribbon, garter, and rare copperbelly water snakes in crayfish burrows. At the Sloughs Wildlife Management area in western Kentucky, MacGregor discovered that copperbelly water snakes were using crayfish burrows present in the dams of constructed wetlands.[60]

Restoring and creating wetlands can benefit crayfish by providing them with habitat where they can reach their highest densities. Crayfish typically colonize locations that contain standing water and areas with saturated soils, bypassing sites that are dry or compacted on the surface. Locales that contain standing water, a high water table, moist soils, and few, if any, trees appear to provide primary and secondary species of burrowing crayfish with optimum habitat. A beaver pond is an excellent example of the type of wetland where a variety of crayfish can be found. Some will use open waters, while others settle on the margins where soils are saturated.[61]

An abundance of cover will increase crayfish survival, and appears to be especially important for young crayfish.[62] Vegetation, branches, and logs can provide the hiding places crayfish need to avoid being eaten by fish and other predators.

It is uncommon to find crayfish and their burrows in forested areas. Apparently, the presence of tree roots and dry soils restricts their ability to dig into the water table. However, crayfish burrows are occasionally found in forested wetlands where trees are widely spaced, soils are moist, and there is standing water.[63]

The eastern box turtle will burrow into the mud surrounding a wetland during the summer to keep cool.

Obtaining Government Permits

Chapter 10

Permits from government agencies may be required before you can legally restore, create, and/or enhance a wetland. These permits may be needed regardless of whether the land is owned by an individual, business, or government agency. However, some of these approvals are necessary only when federal funds are used to pay for all or part of a project.

"Call Before You Dig"

It is critical that you phone the "Call Before You Dig" number before building a wetland or digging test holes. Failure to avoid buried utilities can result in steep fines, and the possibility of injury or death. Expect buried electric, gas, phone, or water lines to be anywhere, even on public land in remote areas. Each region has one phone number to call to notify utility companies of your planned excavation. It is best to make the call early in your preparations so that buried utilities can be avoided with the design. Have available the location of the project, along with the nearest street address, when you call. The project area should be marked so that technicians know where digging could take place. Each company with utilities in the area has a certain amount of time, sometimes up to five business days, to respond. The companies that visit the site will mark underground utilities using the American Public Works Association (AWPA) color code.

National Environmental Protection Act

Projects that use federal funds or that take place on federal lands are subject to the requirements of the National Environmental Policy Act (NEPA). NEPA requires that federal agencies consider environmental values and public concerns in their decision-making process. The steps taken to comply with NEPA for a planned project are generally written in an *Environmental Assessment (EA)* or an *Environmental Impact Statement (EIS)*. It is also important to address the requirements of Section 106 of the National Historic Preservation Act and Section 7 of the Endangered Species Act in the NEPA documentation.

This natural ephemeral wetland is in the Clifty Wilderness on the Daniel Boone National Forest in Kentucky.

Preparing an *EA* or an *EIS* can be a time-consuming job, requiring frequent contact with personnel from agencies involved with the project. Therefore, it is best to obtain an example of the environmental document the lead agency would like you to use before writing an *EA* or an *EIS*. Fortunately, some agencies allow wetland or stream restoration projects to be categorically excluded from documentation in an *EA* or an *EIS*, allowing a much simpler *Decision Memo* to be used.

Section 404 Permits

A *404 Permit* may be needed from the U.S. Army Corps of Engineers and your State Division of Water before building a wetland to comply with Section 404 of the Federal Water Pollution Control Act of 1972, as amended (commonly called the Clean Water Act). This requirement can apply to any action that may affect waters of the United States, regardless of whether or not federal funds are used.

Not every wetland or stream project requires a *404 Permit*. If your project does not affect a jurisdictional stream or wetland, you may not need a permit. Designing a project on dry ground away from streams or existing wetlands is a good way to avoid delays associated with obtaining a *Section 404 Permit*. If the area being considered is poorly drained farmland, it may be best to continue farming operations until the project is done. A *Section 404 Permit* may be needed if agricultural practices cease and hydric plants begin growing naturally on the site.

There are two kinds of wetland when it comes to permitting, those that are jurisdictional, and those that are not. Jurisdictional wetlands have water present during part of the year, hydric soils and hydric vegetation. They are also hydrologically connected to another wetland or stream. If the area where your project is planned contains some standing water and aquatic plants, if water is present in the ground, and the color of the soil near the surface is gray or black, you may be in a jurisdictional wetland, which will require a *Section 404 Permit* in order to proceed. Even a drainage ditch can qualify as a jurisdictional wetland.

The consequences of not obtaining a permit when one is needed may involve having to restore the site to its previous condition, completing off-site mitigation, and being penalized with a fine.

If you work in partnership with a government agency to build the wetland, they will probably handle *Section 404 Permit* needs. And, if you're unsure that an area contains wetlands, *The Wetland Delineation Manual*[64] can help you determine if portions of your project area are wetlands. This manual is available online by visiting most Army Corps of Engineer websites.

Before you assume that permits are needed, look on the internet for the phone number of the nearest U.S. Army Corps of Engineers Office, Regulatory Branch, and call them to discuss the project. However, before you do, it's best to prepare by having the following information ready:

1. State, county, and closest named community to where the project is located
2. Purpose of the project: for enhancement, or as a business venture for future mitigation credit
3. Government agency you are working with to complete the project, if applicable
4. Notes describing signs of historic drainage activities, such as ditches, buried drain lines, lands plowing pattern, leveling, filling, surface inlets, moved and channeled creeks
5. Description of the soil as hydric, if so, by the NRCS Soil Survey that covers the county, and whether the NRCS has labeled it as "PC" for prior conversion
6. Map showing the boundary of your project worksite, ready to be faxed if requested

Be upfront and honest about why you want the permit, as intent may trigger the need for certain kinds of information. Hopefully, your wetland project may be placed on a fast track for approval because of the agency's interest in restoring wetlands.

The information you share may be enough for the agency to determine if a permit is needed. The U.S. Army Corps of Engineers *ENG Form 4345 Application of a Department of the Army Permit* is used for this purpose. You are not required to hire a consultant to prepare this application, but you must provide the Army Corps with the information they need on the form in order to process your application. Beware that the Army Corps may verify the information you provide by field-checking your project area. You will likely have to address the requirements of Section 106 of the National Historic Preservation Act and Section 7 of the Endangered Species Act with your application.

Should the agency suggest you apply for a permit, you may have set the stage for approval to be granted under the authority of *Nationwide Permit Number 27(NWP)*, which covers Stream and Wetland Restoration Activities.

There are two main advantages in designing your project so that it can qualify under *Nationwide Permit 27*. Initially, you'll receive approval to complete the work faster because it won't be necessary to apply for an *Individual Permit* (which involves a 30-day review period). Secondly, expensive mitigation work will not be required.

The *NWP* is designed to cover "Activities in waters of the United States associated with the restoration of former waters, the enhancement of degraded tidal and non-tidal wetlands and riparian areas, the creation of tidal and non-tidal wetlands and riparian areas, and the restoration and enhancement of non-tidal streams and non-tidal open water areas ...." The *NWP* covers activities such as the removal of accumulated sediments; installation, removal, and maintenance of small water control structures, dikes, and berms; the backfilling of artificial channels and drainage ditches; removal of existing drainage structures; construction of small nesting islands and open water areas; planting of appropriate wetland species; mechanized land clearing to remove non-native invasive, exotic or nuisance vegetation; and other related activities.[65]

The *NWP* does not authorize the conversion of natural wetlands to another aquatic use, such as creation of waterfowl impoundments where a forested wetland previously existed. However, the *NWP* "authorizes the relocation of non-tidal waters, including non-tidal wetlands, on the project site provided there are net gains in aquatic resource functions and values."[66]

This language in *Nationwide Permit 27* does not prevent you from changing one type of wetland to another type, for instance a wet-meadow to an emergent wetland. It does indicate that your project will not qualify if you change a natural wetland to an artificial wetland. Therefore, where you believe it to be the case, it is imperative that you make clear you are not changing a natural wetland, but are

The stream side salamander will benefit from wetlands and streams restored in a floodplain. (Cynthia Brookshire photo)

10.Restoring native plants

One way to demonstrate your concern in returning native plants to the wetland you are building is to let the government representatives know that you are purposely leaving portions of older, human- created wetlands intact in order to help supply aquatic plants and animals into the wetland that is being restored. This may involve leaving short sections of ditches that contain wetland features untouched and small areas where drainage lines are not functioning undisturbed. However, be careful during construction to modify these features so that they don't continue to operate after the project is complete.

Activities such as the creation of a "pond" or an "impoundment" are not eligible under the *NWP* and may require that you apply for an *Individual Permit* from the Army Corps, wherein expensive mitigation may be requisite for approval. Being clear in how you describe the intent of your wetland restoration project can reduce the possibility of its being confused with ineligible practices.

restoring an artificial wetland resulting from historic modification to a natural wetland. You must be prepared to explain why you believe existing wetlands in the project area are artificial, resulting from human activities by describing drainage practices observed on the ground.

Your wetland project may be delayed or not approved under the *NWP* if the agency believes that wetlands on the project site are natural wetlands. Never assume that the government person or consultant you are working with will recognize the signs of previous drainage activities within or near the project area. It is best to identify these historic drainage practices in advance, documenting them for the reviewer. However, do be diplomatic in your discussions with these personnel, as they may be wetland experts.

You can also facilitate the approval process under the *NWP* by pointing out "a net increase in aquatic resource functions and values in the project area"[67] realized by implementing your project. These functions and values can include:

1. Providing habitat for endangered and threatened species
2. Improving habitat for migratory birds
3. Restoring spawning habitat for fish
4. Recharging groundwater
5. Reducing potential for flooding
6. Cleaning runoff before it reaches streams, rivers, and lakes
7. Increasing opportunities for environmental education
8. Creating new wildlife viewing opportunities for the public
9. Controlling non-native invasive plants

Section 401 Permit

Section 401 of the Clean Water Act requires an applicant for a federal license or permit to conduct any activity that may result in a discharge of a pollutant into waters of the United States to obtain a certification from the state in which the discharge originates or would originate.[68] A *Section 401 Permit* may be needed from your state if the project affects a steam shown on a topographic map, or one found on the ground, or if the project affects waters flowing in a ditch.

It is wise to contact the agency in advance to discuss the project to determine if a *Section 401 Permit* is needed. Some important information to have available is the length of the stream that may be impacted, and whether the flow regime of the stream is perennial, intermittent, or ephemeral. Should the agency suggest you apply for a permit, work to set the stage for approval to be granted under the authority of *NWP*. Compensatory mitigation is not required for activities authorized by the *NWP*, similar to Section 404.[69]

Activities authorized by the *NWP* include the installation of current deflectors; enhancement, restoration, or creation of riffle and pool stream structure; the placement of in-stream habitat structures; modifications of the stream bed and/or banks to restore or create stream meanders; in addition to activities described by Section 404. The *NWP* does not authorize the conversion of a stream to another

Wood mulch, left over from land clearing operations, can provide an inexpensive and effective sediment barrier.

aquatic use, such as the creation of an impoundment for waterfowl habitat.[70]

Floodplain Permit

A *Floodplain Permit* may be required if your project is planned within the 100-year floodplain as shown on a national flood insurance map. The best way to determine if this permit is needed is to contact a local government floodplain manager. A majority of counties employ an individual who is assigned floodplain management duties.

The main reasons a *Floodplain Permit* is required are to help insure your actions do not affect neighbors, neighboring property, or improvements such as bridges and roads. Your chances of obtaining a *Floodplain Permit* should be increased by considering the following factors in your design:

1. Avoid building high dams.
2. Avoid building dams with steep slopes.
3. Avoid building dams perpendicular to the flow of flood waters.
4. Lower the profile of the wetland by doing more excavation and less dam building.
5. Spread excavated soils in low places down slope from the dam being built, or in low ridges paralleling the flow of flood waters.
6. Move excavated soils to higher ground above the floodplain.

Storm Water Permits

The U.S. Environmental Protection Agency National Pollutant Discharge Elimination System and Permitting Program are managed in partnership with EPA regional offices, states, tribes, and other stakeholders.[71] Each state manages a storm water management program as part of this federal program. Wetland and stream restoration activities can be subject to this program because of the potential for erosion during construction. It is necessary in many states to apply for a *Storm Water Permit* before undertaking wetland and stream construction activities.

What is often the case is that a *Storm Water Management Plan* is required as part of the application for a *Storm Water Permit*. In this plan, it is important to describe the actions being taken to prevent pollutants that can include sediments from leaving the site. Mentioning how the shallow basins that are being constructed for wetlands will trap sediment is a good thing to do.

There are a number of actions that can be taken to reduce erosion during the construction of wetlands and help set the stage for permit approval. To trap runoff, the organic materials and topsoil that are removed in the early stages of construction can be wind rowed along the lower edge of the work area. These materials can later be spread over the wetland in the final stages of construction. Shallow depressions can be dug along the lower edge of the construction site to collect runoff. These sediment basins can be reshaped and included in the final design of the wetland.

To prevent sediment-laden waters from flowing off-site, silt barriers consisting of filter fabric fences and staked straw bales can be placed along the lower edge of the work area. Wood chips left over from right-of-way clearing operations can also be used to form an inexpensive, effective, and environmentally-friendly sediment barrier. The piles can simply be spread once the exposed soils are growing vegetation.

Loosening the top 18 inches of soil over a completed wetland project is an excellent way to prevent soils from washing. The bucket of an excavator can be used to turn over compacted soils, working the equipment backwards, much like washing a floor. The loosened soils also provide a superior substrate for wheat, grasses, wildflowers, trees, and shrubs to grow. To further prevent erosion, the exposed soils should be seeded to wheat and mulched with straw the same day the wetland is completed, as described in Chapter 8.

Erosion is more likely to occur along stream banks and wetland spillways, where water is flowing, rather than in the wetlands themselves that hold standing water. A mesh erosion control blanket made from coconut fiber can be placed over straw and anchored to prevent erosion in these areas. The heavier, 400-lb material containing larger openings works well to hold soils in place. Erosion control blankets containing plastic mesh should be avoided because they can entangle wildlife.

National Historic Preservation Act

Projects on federal land and those on state, county, and private lands receiving federal funding must comply with Section 106 of the National Historic Preservation Act of 1966 as amended. This Act established a program for the preservation of additional historic properties throughout the nation, along with other provisions. The Act directed federal agencies to provide leadership in the preservation of the prehistoric and historic resources of the United States, the international community of nations, and in the administration of the national preservation program in partnership with states, Indian tribes, native Hawaiians, and local governments.

When federal funds or lands are involved, you need to make sure that your project does not affect sites containing artifacts more than 50 years old without prior written approval from the State Historic Preservation Officer (SHPO). Archeologists typically complete a *Cultural Resource Survey* of the proposed wetland project, which involves examination of the project area for artifacts, and the preparation of a report. The report makes recommendations for possible concurrence by SHPO regarding whether or not implementing the project would affect sites that are listed or are eligible for listing on the National Register of Historic Places.

A federal agency may be willing to complete the *Cultural Resource Survey* for a wetland project planned on private land. The NRCS and U.S. Fish and Wildlife Service are examples of federal agencies that will partner with private landowners to restore wetlands by completing Section 106 requirements.

Cultural Resource Surveys can be very expensive to complete, and in some situations their cost can exceed wetland construction costs. They can take a lengthy time to complete; waiting one year for their results is not uncommon. Therefore, it is critical to design your wetland project with the protection of cultural resources in mind. Employing the following actions will greatly reduce the time and expense it takes to comply with the 1966 Act.

Inform the archeologist conducting the survey that you plan to design the wetland project to avoid impacting cultural resources. Your goal should be to make it possible for the archeologist to prepare a "No Find" report, which will save time and money. Avoiding cultural resources is the fastest and cheapest way to have the survey completed. This means that you must be willing to abandon construction sites or portions of construction sites where cultural resources are found. Anticipate this happening and have alternate work locations available for the archeologist to examine at the time of the survey.

Ask the archeologist to mark the boundary of any and all cultural resource sites found within or near the planned project. Since most cultural resource sites are small, being under one acre in size, you should be able to design around them. It is generally cheaper to move more soil, and to move this soil a greater distance than it is to complete further archeological analysis of the alternate sites you may want to use.

When planning a wetland project, it is best to mark each wetland site as close to the edge of the planned disturbance area as possible. Colored flagging hung from trees and shrubs, and plastic wire flags set in the ground work well for this purpose. Requesting a survey of larger areas will greatly increase costs. Money can be saved by designing smaller wetland projects, and then marking the area to be altered before the *Cultural Resource Survey* is begun.

Some ways of reducing costs and the possibility of including a cultural resource site in a wetland project area are:

1. Select low and wet areas for your project, since most archeological sites are on higher ground.
2. Incorporate only lower terraces when laying out a project along a river or large stream unless it's absolutely necessary. It's on these second or third terraces above the water where people lived.
3. Avoid including old home sites in your area; these may be cultural resource sites that require documentation, and there is higher than a fifty percent chance that they're also pre-historic sites.
4. Flag areas that are on the slope, when on a ridge, as cultural resource sites are generally found on flatter areas. The added cost of moving more soil on a sloped area can be offset by lower reporting costs.
5. Build access roads off to the side of a ridge and along the slope to avoid hitting cultural resource sites on flatter ridge-tops.
6. Keep the blade up on the dozer and weave between the trees when traversing equipment through the woods to avoid the need to survey access routes.
7. Prepare high quality maps showing the location of the proposed project in an electronic format to speed the preparation of the report.
8. Mark each proposed project area with flags in the field so that the areas can be easily found by the person conducting the survey, saving his time and helping to reduce costs.

Avoiding old home sites, which include the house, barn, and outbuildings, is an easy way to keep costs down on a wetland project. Often marked on old topographic maps, home sites are readily identified by looking for the following features when marking the boundary of the area where ground disturbance is planned:

1. Chimney, cement foundation, and steps
2. Large, flat stones which were used for footers

3. Raised ground with large trees and different grasses
4. Fruit trees present (apple, cherry, pear, and plum)
5. Ornamental flowers, such as daffodils, iris, and yucca
6. Household and/or farm debris strewn around
7. Dark-colored soil and gravels
8. Utility poles and cables

Added expenses can be incurred when planning wetlands in a major river valley, due to the potential for finding buried soil horizons, which may contain cultural resource sites. Deep testing with a backhoe may be required to determine if alluvium is covering artifacts.

It costs money to legally impact cultural resource sites identified by the *Phase 1 Cultural Resource Survey*. A *Phase 2 Cultural Resource Survey* must be completed for each site to be disturbed by the project. The purpose of the *Phase 2* is to determine if a site is eligible for listing on the National Register of Historic Places. One can expect to pay at least $10,000 for each site subjected to a *Phase 2 Survey*. If the *Phase 2 Survey* finds that the site is eligible for listing, and you still want to proceed, a *Phase 3 Survey* will be required, which involves the excavation of the site. Expect to pay at least $50,000 for each of these surveys.

Once your project is approved, it is important to ensure that neighboring cultural resource sites are not affected by heavy equipment. Cultural resource sites should be placed on contract maps and clearly marked in the field with bright flagging of a different color than that used to mark areas to be cleared. Strongly warn each equipment operator not to disturb cultural resource sites, or he may be personally liable for a large fine. Attention to detail is especially important when moving equipment onto a worksite, and during the clearing operation, as this is when violations are most likely to occur.

Endangered Species

Projects completed on federal land and those on state, county, and private lands receiving federal funding must comply with Section 7 of the Endangered Species Act. It is a legal requirement that federal actions do not harm endangered or threatened species, or their habitats. The Endangered Species Act, passed in 1973, was designed to help prevent certain populations of plants, animals, and invertebrates from becoming extinct. Species can be listed as "endangered" or "threatened" by the U.S. Fish and Wildlife Service, and land important to their survival, called "critical habitat", may also be designated by the agency.

The U.S. Fish and Wildlife Service has the responsibility of ensuring that actions funded in whole or in part by the federal government, including wetland and stream restoration, will not harm listed species, or negatively affect their habitats. However, in order for the U.S. Fish and Wildlife Service to make this determination about a proposed project, information must be made available to them, generally documented in a report prepared by the proponent, known as a *Biological Assessment*. Once agency personnel decide that their qualifications have been met, a letter is issued to the person proposing the activity to the effect that the projection, as described, has been approved.

It's best to talk with an agency employee before you finalize the design for a project, so they can help determine if listed species are present in the area, and identify measures necessary for avoiding harm to these species. To aid you in preparing a *Biological Assessment*, the agency can provide you with a list of endangered and threatened species found in your area. They can also help you plan how to protect listed species during construction.

You may find that the U.S. Fish and Wildlife Service will want to assist you with your wetland project, especially since most can be designed to benefit endangered and threatened species. Because over one half of all listed species use wetlands, it may be possible to plan your wetland project to benefit one or more endangered species. In addition, the U.S. Fish and Wildlife Service, like all other federal agencies, have been directed to reverse the downward trend in wetland acreage, and working in partnership can contribute to meeting this goal.[72]

The US Fish and Wildlife Service encourages private landowners to build wetlands, and looks for partnerships to restore wetland and stream habitats. You may want to ask if they would be willing to prepare a *Biological Assessment* for you at no charge in order to help complete the project.

Renovation and Maintenance

Chapter 11

Spending money to maintain wetlands can be a wise investment. For what is generally less than the cost of building a new wetland, poorly designed, nonfunctioning, or damaged wetlands can be refurbished. Wetlands constructed years ago may develop simple problems that, when repaired, can be expected to last for generations.

Convincing someone to spend money on a failed wetland can be more difficult than selling a car with body damage; people do not believe that something that looks so bad can contain good inside. A failed wetland has a stigma attached to it; just as no one brags about purchasing a vehicle with a bad engine, it is rare to be shown a wetland that will not support cattails. Fortunately, just as with vehicles, failed wetlands can be fixed.

There are many good reasons to repair a wetland. Quite often, land is limited for wetland restoration, and it is better to finance the repair than to find another site for construction. Or, maybe the failed wetland is an eyesore visible to the public, and it is an embarrassment to an agency or a landowner. Perhaps the wetland could help endangered species, or it really is a good place for environmental education. But, in reality, the best motivation is growing tired of hearing comments about what one has failed to accomplish.

Repairing a wetland is similar to the steps an honest mechanic takes when fixing a car: he does not replace the engine when only a plug wire is loose. There are things that can be checked to help determine why a wetland will not hold water, starting with the easiest and cheapest, then moving on to the more difficult.

It is always worth talking to the person(s) who designed the failed wetland, even if they are retired. Just as importantly, speak to the heavy equipment operator who was out there pushing soil. It will be surprising how much this person will remember, usually because of the novelty of building wetlands in his line of work. Time spent in the role of detective can alert one to what to look for when examining the site on the ground.

The best way to identify what might be wrong with a wetland is to walk around it several times. Try to determine if a dam was built for the wetland. The dam is generally a long ridge of dry, compacted soil with water on one side. If soils in the bottom of the wetland are high in clay, then a dam was used to build the wetland. Locate the dam and walk the entire length. Pay close attention to the top and backside of the dam.

- Are there any holes or ditches in the dam? These can carry a lot of water.
- Are trees and shrubs growing on the dam? Their roots are like small drainpipes.
- Is the dam rough on top? It may have been built from topsoil that leaks.
- Does the dam have steep sides? Such a dam may lack compaction, and is easily damaged by muskrats.

Walk the backside base of the dam. If one finds flowing water, wet spots, and patches of aquatic plants, then water is leaking under the dam. Walk downhill from the dam and look for flowing water or evidence of a small stream channel. If water is flowing, it is probably coming from the wetland. Topsoil, buried drain lines, gravel, or wood left under a dam can provide a path for waters to flow from the wetland and under the dam.

Examine the elevation of water in the wetland. Unless it is a dry time of year, it should be near the top of the dam. Determine if the water level is at design levels. Examining a wetland for problems in late summer can provide conflicting results. If the wetland was built shallow, it may be normal for it to dry in late summer or fall.

- If the wetland was known to hold water originally, look for muskrat and beaver holes in the dam, or a faulty water control structure.
- If the water level drops within three days of a heavy rain, buried drain lines are probably present.
- If water levels drop slowly, topsoil may have been left under the dam.

Verify if a water control structure or drainpipe is present. These are generally located in the deepest part of the wetland, much like the drain in a swimming pool. Not every wetland has a water control structure, but if it does, it will be located along the dam where the deepest water is expected to be. A broken, vandalized, or leaky water control structure will ruin the hydrology of a wetland. Many wetlands can be repaired by simply closing drainpipes inadvertently left open years ago. Failure can also occur because holes are found in the overflow pipe.

Finding a flashboard riser can explain why the wetland failed. These vertical steel culverts contain a channel for dropping down boards to adjust water levels. The flashboard riser is attached to the end of a culvert that passes beneath the dam. By their nature they leak, and are almost impossible to prevent from leaking. This may not be the problem if a perennial stream flows into the wetland, but they are a problem if the wetland has a small watershed. Remove the flashboard riser and its culvert, replacing it with a PVC drainpipe and plastic water control structure.

The gaskets which seal the plates in fiberglass and plastic water control structures can begin to leak over time. This problem may occur in wetlands that have held water for years, but become dry unexpectedly. The gaskets which seal the plates within these structures may compress or tear, allowing waters to seep past them. The flow past the gaskets is greatest after a heavy rain when the wetland is filled, but is barely discernable after the water level has dropped. The plates on these water control structures should require some strength to slide in and out, so when they wiggle or move easily, they must be replaced.

A drainpipe and water control structure can be hard to find if the site has grown up to trees and shrubs. If a water control structure is not found, look for a drainpipe. The presence of a ditch along the backside of the dam can indicate where the drainpipe is located. Examine the outlet for flowing water; it should be dry. Look for eroded ditches and bent vegetation below the drainpipe that show where water is flowing from the wetland after a rain. If the outlet is covered with mud or blocked by beaver, dig into the obstruction to see if water seeps from the end. A trickle of water from the end of the pipe on a dry day can be a strong flow after rain. Water pressure will exacerbate the leak and cause waters to rapidly leave the wetland.

To check the quality of the water control structure, confirm the following:

- Was it left open? Someone could have forgotten to replace the plug.
- Is it leaking? Carefully examine the seals to find if they leak or are loose.
- Is there evidence of damage? Look for bullet holes.
- Is water flowing in the drainpipe? Drop in crumbled leaves and stir up mud to see if waters are moving in the pipe.

It is also important to inspect the path of waters flowing from natural and constructed wetlands. Action is needed if the stream contains signs of erosion, or if one or more head-cuts have formed. A head-cut will advance upstream into the wetland, acting like a ditch to remove standing water, and lowering the elevation of groundwater in the wetland. Head-cuts are the primary factor responsible for passively draining wetlands over time.

Most wetlands have a spillway, or a low place at one end of the dam. The spillway is a wide excavated channel, lower than the top of the dam that provides a path for water to leave the wetland without causing damage to the dam. Spillways are generally constructed across undisturbed soil, may be rocked, and should have gentle slopes. Check to see if the spillway is blocked by leaves, branches, soil, or by a beaver dam. Remove any material restricting the spillway to restore its capacity to carry excess waters from the wetland.

The last step in inspecting a failed wetland involves examining the bottom of the wetland.

- Is gravel present? Scattered creek rock on the surface

This natural wetland on the Monongahela National Forest in West Virginia is being drained by head-cuts moving upstream through the wetland. The head-cuts began when a culvert was placed in a stream crossing downhill from the wetland. The head-cuts have formed a channel that is moving uphill through the wetland, draining surface water and lowering the elevation of groundwater in the wetland.

A head-cut has advanced upstream into the spillway of this constructed wetland and will soon drain the wetland.

indicates a permeable layer that is buried below.

- Are the soils in the dam and in the bottom of the wetland high in clay? If not, waters may be simply soaking into the ground.
- Are there any vertical holes six or more inches in diameter? These may be blowouts that indicate missed buried drain lines.
- Are crayfish burrows present? If yes, is water near the surface in the burrows? If the holes are dry, water is probably flowing down the holes and under the dam via a subsurface permeable layer.
- Is there a liner? Finding one after digging a test hole by hand indicates it was damaged during construction, and is now leaking.

In many situations, the soil samples taken in the bottom of the wetland contain enough clay that a ribbon almost two inches long can be formed without breaking. While these soils should hold surface water, they are not. This indicates that water is probably soaking into the ground, and quite possibly, traveling under the dam. Water can leave a wetland by finding its way into buried drainage structures, by passing through a topsoil layer, or by following a buried layer of sand and gravel. Each one of these problems can be fixed by building a groundwater dam. Digging a test hole with an excavator can help determine the size of the job.

Walk around the bottom of the wetland and look for holes in the ground. A crayfish hole may only be one inch in diameter, and its entrance can be covered by a pile of soil or a chimney that sticks above the ground. Crayfish holes go down to water, even if it is more than 20 feet deep. Surface water can be trapped in the hole like a mini-pond, or the bottom of the hole can end in groundwater that is flowing through a layer of sand or gravel. Either way, if crayfish burrows were left intact under the dam while it was built, they can provide a path for waters to leave the wetland. Pouring water into a crayfish burrow can show whether permeable soils extend beneath the dam. If the waters poured from a bucket disappear into a hole, one can assume that the soils below are permeable. Coring the dam will crush and block these burrows, and will also interrupt permeable layers beneath the dam, effectively repairing the wetland.

Sometimes, larger holes or sinkholes may be found in the bottom of a wetland with evidence that surface waters are flowing into the ground. These holes often indicate where waters are directly entering a buried system of drain lines, or even a limestone cave. Again, a groundwater dam should be constructed to block the underground movement of water beneath the dam.

Should buried clay tiles or plastic drain lines be present under the wetland, it may be possible to find their outlet. Walk along the creek, river, or ditches alongside and up to one-quarter mile downstream of the wetland. Look for gaps in the ditches or along the creek bank. If one spots a pipe, it is probably an outlet. Sometimes, one is fortunate enough to see water flowing from the pipe, but typically, there is only a small wet spot that is easily missed.

This failed wetland on the Fleming County Wildlife Management Area in Kentucky has several small pools of water near the dam. (Wes Mattox photo)

A vertical hole 12 inches in diameter was found in the bottom of the wetland containing these clay tiles. Water could be heard flowing underground, proving the tiles were still doing the job they were designed to do. (Wes Mattox photo)

This recently cleaned crayfish burrow was found in a drained emergent wetland. (Dennis Eger photo)

This outlet for buried clay drain tiles empties into the Licking River in Bath County, Kentucky. The outlet looks like a short, dead-end ditch along the riverbank.

A few pieces of clay drain tile are sometimes visible near the end of an outlet.

This outlet for the buried drainage system is difficult to see in the deep ditch. Suspect that drain lines were buried in the area if ditches and straight streams are found within one mile down slope. (Barry Nichols photo)

The outlet for this buried drainage system empties into a roadside ditch on John D. Smith's farm in Menifee County, Kentucky. The clay pipes continue to function even when the outlet is covered with soil.

Once the outlet has been found, it is then necessary to dig back along the line, through the dam, and into the wetland area to expose and block the line. However, it is also possible that there are more drain lines, and that the outlet for only one of these systems has been discovered. Should wood or rock have been used to drain the wetland, there is little chance one will find the outlet. A groundwater dam should be constructed along the entire length of the dam to block these.

Subsurface Sand and Gravel Layers

Wetlands built from silty-clay-loam soils have failed in Kentucky, Illinois, Indiana, Maryland, and Minnesota because these fine-textured soils are found on top of a permeable layer of sand and gravel. Surface waters will flow into crayfish burrows, through these impermeable soils until they reach a sand or gravel layer, join with the groundwater, and then leave the wetland by traveling under the dam. The permeable layer may be from two to 20 feet below the surface. The subsurface permeable layer can be blocked by using a dozer and an excavator to construct a groundwater dam. Certain steps should be taken to confirm the presence of a buried permeable layer before the job is undertaken.

Carefully examine the soil around the entrance of crayfish burrows in the bottom of the wetland to find if it contains small rocks. The presence of gravels can indicate a permeable layer deep in the ground. Walk downhill from the wetland to the nearest main creek or river and find a vertical cut bank. If there is a layer of sand or gravel there, then the permeable feature probably extends under the wetland area. One method for determining the thickness of a gravel layer beneath a failed dam involves subtracting

A large dozer and an excavator build a groundwater dam that prevents water from flowing under the dam. The excavator is removing a buried plastic drain line.

An excavator removes eight inches of soil from the bottom of a failed wetland in preparation for installation of a liner.

Soils are replaced over the liner that was sandwiched between two, 8-ounce geo-textile layers.

the elevation of the top of the gravel layer where the dam is located from the elevation of bedrock exposed in the creek. The elevation of the bedrock in the creek is usually about the same as the elevation of bedrock beneath the dam. In most cases, water will be flowing over the bedrock layer beneath the dam into the creek below. Failing to find exposed patches of bedrock in the creek opposite the constructed wetland may indicate a very thick layer of gravel under the failed wetland and dam.

The most reliable way to find a buried permeable layer is to dig test holes in the bottom of the wetland with an excavator. Ask the equipment operator to dig a series of deep holes along the inside base of the dam. Draw a sketch of each test hole that shows the thickness and depth of each soil layer. This information can help determine if it is feasible to repair the wetland, and to prepare a contract.

Test soils in the constructed dam to see if it was built from suitable material. The middle of the dam will need to be replaced if it was constructed from topsoil, gravel, or sand.

In order to repair a wetland with a buried permeable layer, there must be a readily available source of soil high in clay that is close to the dam to replace the gravel found beneath the dam. To verify this, it is important to dig additional test holes up to 300 feet from the dam to see if clay soils are accessible. It works best to have a layer of fine-textured soil that is at least three feet deep over the bottom of the wetland that is available for placement in the groundwater dam. Should this soil not be close at hand, it can be hauled by truck or by scraper, but this can add considerably to the cost.

Providing the dam is constructed from clay, the wetland can be repaired by building a groundwater dam beneath the inside slope of the existing dam. The inside slope of the dam is moved out of the way, a groundwater dam is constructed beneath its location, then the inside slope is reshaped over the completed groundwater dam.

Installing a Liner

There are situations where the most feasible way to repair a wetland is to install a liner. This can be the case where soils are low in clay, or where it is not possible to dig deeply enough to reach the water table, disable drain lines, or build a groundwater dam. Installing a liner is a reliable way to repair a wetland, but they are expensive, and usually are not practical for wetlands greater than 3,600 square feet.

Muddy Water

Check to see if water is muddy in the wetland. Finding muddy water and few aquatic plants, combined with evidence of fishing, indicate the wetland contains fish, such as non-native carp, and should be drained to return plant diversity and wildlife use.

Reducing Maintenance

Wetlands restored or constructed years ago that have required no maintenance share a number of characteristics that can be useful when building or maintaining other wetlands:

- The wetland is isolated, in that rainfall and runoff entering the wetland stays in the wetland.
- The wetland has a spillway that is wide-at least 12 feet or wider if the wetland has a small watershed and 40 feet or wider if it has a large watershed. The wide spillway allows water leaving the wetland to follow a path that is so gradual that beaver do not block it with a dam.
- Water leaving the wetland follows a spillway down a slope that is three percent or less. The flow is so gradual that it does not cause erosion or form a head-cut that can drain the wetland.
- If the wetland has a dam, the dam is three feet high or less, has a level top that is 20 feet or wider, and slopes on the inside and backside that are five or less percent.
- The wetland does not have a water control structure or drainpipe.
- The wetland is not adjacent to another wetland or to a stream. This factor is important because of beaver that can dig channels connecting wetlands to other wetlands and to streams. If two wetlands are next to each other, and one is on higher ground, a beaver-dug channel can drain the upper wetland into the lower wetland. If the wetland is close to a stream, water can follow the beaver channel like a ditch to the stream.

At any time when heavy equipment is needed to maintain a wetland, this is an excellent opportunity to modify that wetland so that it will require less maintenance in the future. Modifications can be made to make the wetland appear more natural, and improve conditions for plants and animals. These actions should be considered when heavy equipment is available:

- Fill holes in the dam caused by beaver and muskrats
- Remove trees that are growing on the dam
- Add logs and dead trees to the wetland
- Place more gradual slopes on the front and the backside of the dam
- Widen the spillway and place it on a more gradual slope
- Replace the water control structure if it is leaking or damaged
- Modify the water control structure so it is less likely to be blocked by beaver
- Build islands and peninsulas within the wetland
- Place snags upright in and around the wetland

Raising Dams

When looking for ways to increase wetland acreage, one method often suggested involves raising the height of a dam to flood more ground. Seldom is this a good prospect when compared to building additional wetlands. While raising a dam may flood more ground, it can have serious drawbacks.

Before raising a dam, it is a good idea to ask if those who built the wetland were competent. If they were, then a good question to ask is why they did not build it higher in the first place. One can generally assume that the majority of wetlands constructed were designed to maximize the surface area of water created for the height of the dam that was built.

There are problems often encountered when raising a dam:

- The change floods little additional land at a considerable cost.
- The change creates more deep water, and shifts uphill the shallow waters used by waterfowl, wading birds, amphibians, reptiles, and plants uphill.
- There is great difficulty locating soils high in clay needed to raise the dam as the majority of these soils were used for construction. Clays are often covered or mixed with soils wasted from the original construction.
- The new higher dam will have steep sides and a narrow top, with deeper water in front. The steeper dam will be more costly to maintain, and will be prone to muskrat and beaver damage.

Rebuilding Ponds

Hundreds of thousands of ponds have been built across North America to provide water for livestock, fishing, enhancing the scenic value of property, and for treating runoff. Unfortunately, the majority of these ponds look artificial, and lack the characteristics found in wetlands that benefit plants and animals. Most constructed ponds are deep, have steep sides, mowed edges, and consequently, support little more than cattails and Canada geese. As lands are developed, these ponds are often filled and drained. This is unfortunate, as they can be rebuilt to provide naturally-appearing and attractive wetlands in an urban setting. Often, their history of holding water shows that they can be modified to resemble wetlands with success.

Water flowing over the steep, high dam on this farm pond is causing much erosion. Modifying it by lowering the dam, placing gradual slopes on the dam, and creating a spillway would stop the erosion and give it wetland characteristics.

This sediment pond has steep sides and is shaped like a rectangle.

Trees growing on a steep and high dam may cause failure of the pond. Trees should be removed and the dam lowered to reduce maintenance and give the pond wetland characteristics.

The same pond after an excavator was used to create wetland characteristics including gradual slopes, peninsulas, and shallow bays. (Nina Hemphill photo)

Constructed ponds can be rebuilt so they are safer for individuals to enjoy. Many schools and subdivisions built on farmland contain deep ponds originally constructed to provide water for livestock. These sites are inviting for people to approach, but with their steep sides and sharp drop-offs, can represent a safety concern. In addition, many of these old ponds contain fish, and therefore, support few amphibians, aquatic invertebrates, and plants. Fortunately, these ponds can be modified into wetlands that attract wildlife, provide for environmental education, are pleasing to view, and increase property values. Changing an artificially-appearing pond into a naturally-looking wetland involves these main steps:

- Remove water: Look for a drainpipe that passes through the dam. Should one be present, it can be found in the bottom of a ditch along the backside and center of the dam. The drainpipe may be closed with a valve or cap that can be opened. If a drainpipe cannot be found, use an excavator or a backhoe to drain the pond by digging a deep trench through the dam. The trench must be deep enough to drain the majority of waters in the pond. It is important to take steps before breaching the pond to control erosion. Staking rows of hay and silt fabric across the path of the waters in advance will help trap sediment released by waters passing through the dam.

- Allow time for drying: The bottom of the pond and the dam should be allowed to dry for a month or more before additional steps are taken. Old ponds generally contain a deep layer of mud, which is too soft for heavy equipment to move or reshape. When left exposed to the sun, it will dry, and its organic content will decompose, eventually turning into soil that is firm enough for a dozer to move efficiently. Attempting to rush the process is not wise, as heavy equipment will become hopelessly stuck in the soft sediments.

- Remove trees from the dam: A dozer or an excavator should be used to remove trees and shrubs from the dam. These trees can be saved for later placement in the finished wetland to provide turtles with sunning logs, birds with perches, and to improve breeding conditions for salamanders.

- Lower the height of the dam: Use soils removed to place gradual slopes on the front and back of the dam. Gradual slopes, no steeper than ten percent should be placed on both sides of the dam. These gradual slopes will reduce the danger of people slipping and falling into the water, and provide habitat for a greater diversity of plants. Generally, the soils needed to make gradual slopes on the dam can be obtained by lowering the height of the dam, which also accomplishes another objective in reducing the depth of the wetland. Should it be necessary to maintain the depth and size of the pond, soils can be hauled by truck, or taken from an adjacent area to create gradual slopes on the dam.

- Remove corners and straight lines: Square or rectangular ponds should be given a more natural shape.
- Reduce slopes: Some or all of the steep slopes surrounding the pond should be made more gradual.
- Vary the bottom: The flat and smooth bottom found in most ponds should be changed by excavating shallow dips and creating islands, with some rising above the water and others remaining below the surface.
- Repair the dam: The ditch dug through the dam should be widened to at least the width of the dozer blade, and its bottom scraped clean of woody debris and vegetation with a dozer. It should then be filled and packed with clay in layers to prevent leaking.
- Consider adding a drainpipe: Rebuilding a pond provides an excellent opportunity for installing a drainpipe. Generally, the ditch dug to drain the pond is

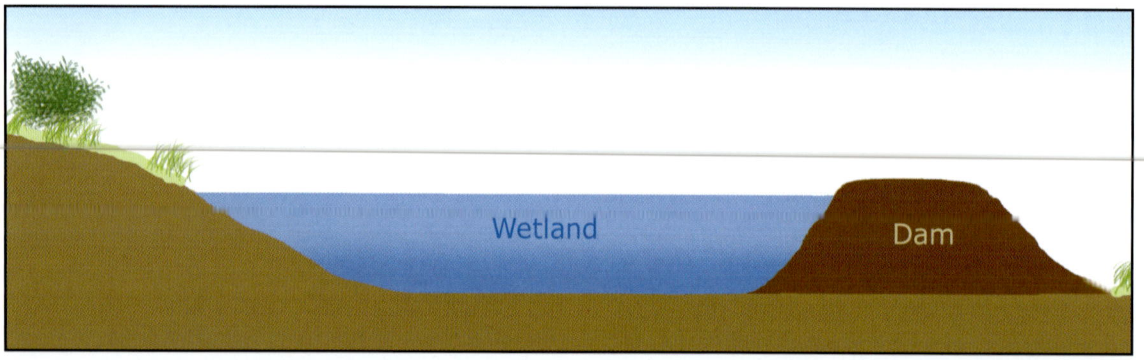

An artificial-looking farm pond, Profile view

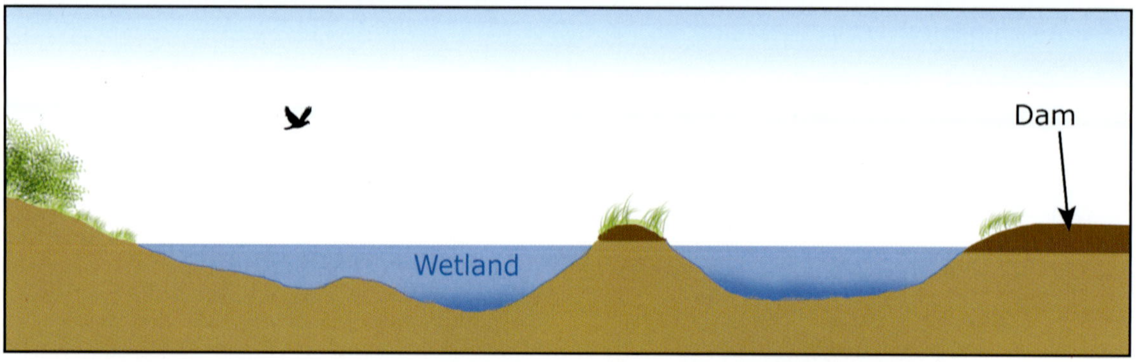

Farm pond rebuilt with wetland characteristics, Profile view

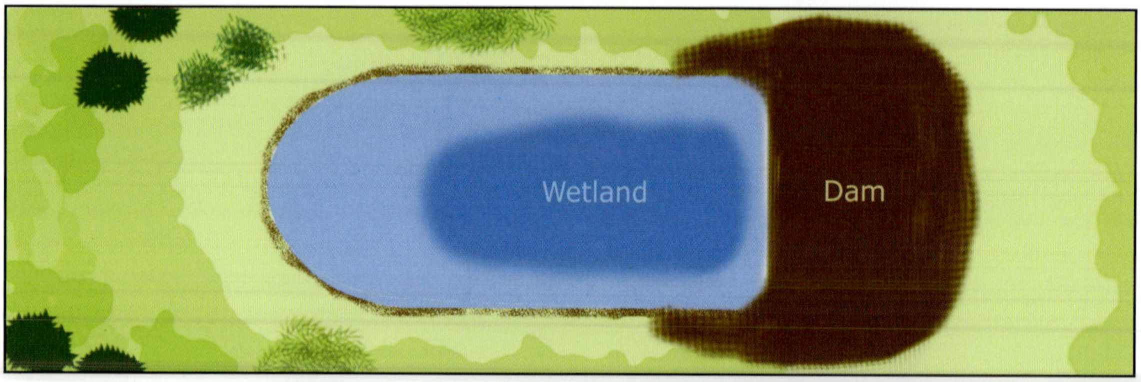

Artificial-looking farm pond, Plan view

Farm pond rebuilt with wetland characteristics, Plan view

A dozer reduces the height of the dam and creates gradual slopes, while an excavator creates peninsulas, inlets, and woody debris, transforming a pond into a wetland.

The dam was made invisible in this pond renovation project.

This photograph of an old pond on Chris and Melissa Yearick's farm in Chemung County, New York shows a steep and high dam that was from two to three feet above the water level in the pond. Muskrats had been tunneling into the dam and causing damage. (Melissa Yearick photo)

The elevation of the dam was lowered 1.5 feet by using a track-steer and a dozer. Gradual slopes were placed on the front and back of the dam. Peninsulas, bays, and islands were added. Exposed soils were planted to native wetland species. A second wetland is being built uphill from the pond. (Melissa Yearick photo)

The renovated wetlands are now providing improved habitat to wildlife and plants. Muskrats are tunneling into the islands, and not the dam. Toad houses and wood duck nest boxes were placed around the new wetlands. (Melissa Yearick photo)

where a new drainpipe will be placed. The bottom of the ditch should be cleaned and packed with a dozer to prepare a firm foundation for the drainpipe. It should be placed in the bottom of the ditch, so that the majority of water in the wetland can be removed. The inlet of the pipe should extend into the wetland pool area far enough so that a gradual slope may be placed over the pipe on the inside of the dam. If it is decided to add a drainpipe to the pond, it will be necessary to extend both the inlet and outlet to place gradual slopes on the dam. Care must be taken when covering the new drainpipe. Soils high in clay should be placed in layers from six to eight inches thick over the pipe, and each layer compacted by the dozer. Waters will leak along a drainpipe surrounded by loose soils and soils contaminated by organic material, causing the pond to fail.

- Consider installing a water control structure: A water control structure will allow for the gradual raising or lowering of water levels, important to providing shorebirds with habitat, removing unwanted fish, and drying soils for future repairs.
- Plant native species: The process of changing a pond into a wetland generally results in burying both native and non-native plants, offering a clean slate for establishing native species on both high and low ground. Seed the exposed soils immediately following construction to winter wheat, and mulch with wheat straw to control erosion. Hand-sowing native grasses and wildflowers, such as switch grass, Indian grass, and partridge pea on the dam and areas above the waterline is also a good idea at this time.

Ponds built with help from the NRCS often have a cement trough located downhill for livestock watering. A pipe that was installed beneath the dam carries water to the trough. The valve on this watering trough is leaking, explaining why the adjacent pond constructed years ago seldom holds water.

Mitigation Notes

Wetlands built to meet the requirements for mitigation seldom resemble the natural wetlands they were designed to replace. This change in appearance may explain the general lack of acceptance of mitigation programs across North America. While the government requires that wetlands built for mitigation have hydric soil, hydrophytic plants, and presence of water during the growing season, it does not necessarily require the developer to reshape the land to more natural contours.

Wetlands considered by the government to be successful mitigation sites often do not look like natural wetlands because of these factors:

- Deep and shallow drainage ditches were left in place.
- Land was left sloped towards drainage ditches.
- Diverted and straightened streams were not restored.
- Buried drainage structures were left in place.
- Depressions containing water were not formed.
- Trees and shrubs were planted in rows at even spacing.

Wetlands built for mitigation can look and function like natural wetlands instead of square-shaped, shallow ponds. This can be accomplished by eliminating the signs of historic drainage, removing functioning drainage features, and returning pools of various sizes, shapes, and depths to wetland restoration projects.

Livestock Watering Troughs

The NRCS helped farmers construct tens of thousands of livestock watering ponds in the United States. To protect this investment, these ponds and dams were generally fenced, and cattle were only allowed access to water via a cement watering trough positioned downhill from the dam. A pipe was installed through the dam to carry water underground into the watering trough. These ponds can fail if the valve on the watering trough breaks or is stuck open. Repairing the valve often repairs the pond. However, in many cases, steel pipe was used between the dam and the watering trough, and this steel pipe has since rusted in acid soils. Leaks from these broken pipes can surface, just like a spring. Finding saturated soils and aquatic plants growing in a zone representing a straight line between the watering trough and the dam can indicate that a break has occurred in the waterline.

Holes in the Dam

Beaver-excavated holes are easier to find than muskrat burrows because they are so large. Beaver will dig into a dam to create below-ground living quarters, and excavate channels entirely through the dam to connect the wetland with another wetland or stream.

A muskrat can drain a wetland overnight by burrowing into

Muskrat holes in a dam can be found more easily when waters are removed and the dam is mowed.

the dam. Muskrat holes can be difficult to find when the dam is growing tall, dense grasses along with shrubs and trees, so it is often necessary to mow the dam to make an inspection. If trees on the dam have grown too large for it to be mowed with a tractor and bush-hog, a dozer may be needed to remove them.

Search for holes by walking along the top of the dam, then along the inside edge, and finally along the backside of the dam. Muskrats begin a burrow by tunneling into the dam from beneath the water in the wetland. Their burrow will lead upwards in the dam, ending in an underground living cavity within the dam that is above the water level in the wetland.

One may be able to see collapsed muskrat burrows while walking the top of a dam. In many cases, the below-water entrances to their burrows can be seen while walking the inside edge of the dam. Not every muskrat burrow will cause a leak. Test each burrow to see which of them need repair by muddying the water at the entrance to the burrow, and then watching to see if waters flow into the burrow. Waters flowing into the burrow indicate they are also leaving the wetland.

Holes in a dam can be repaired with a backhoe, dozer, or excavator. Use bright flagging to mark the holes' locations, so they can be found later when heavy equipment is on the site. To repair, scrape vegetation and topsoil off the dam around the hole, removing organic material from the burrow. Dig a wide depression down to the bottom of the burrow, and replace soil in layers, running the equipment over each layer for compaction, ensuring the patch will not leak. Additional soil may be needed to fill the hole, and this soil can often be obtained from the bottom of the wetland. It is best to drain waters in the wetland before repairs are made; otherwise, water from the wetland can turn the site into mud, and mud cannot be compacted.

Wetlands with muskrats can be modified to reduce tunneling into the dam. One easy way is to increase the elevation of water levels in the wetland so that twelve inches or less of the dam remains above the water level.

Raising the water level can be as simple as adjusting a water control structure or raising the elevation on the spillway. Many dams that are plagued with muskrat problems also have steep slopes. These slopes can be made gentler by adding soil to them. Muskrats are less likely to tunnel into a dam if it has an inside slope of 10:1 or more gradual, and if the water level is near the top of the dam.

It may be necessary to use a level and rod to take elevation readings along the top of the dam to see if leveling is needed prior to adjusting water levels in the wetland. Dams may not have been constructed level, with high and low places being found along their length. The high places are where muskrats live. These humps can be scraped off and moved over into the lower places so that water levels can be raised along the dam's entire length.

Catching More Water

In some cases, a dry wetland can be repaired by increasing the amount of water it collects. For instance, a failed wetland at a school may benefit by redirecting waters from a roof gutter, or by changing where a ditch empties into a storm sewer. For larger wetlands, it may be possible to redirect waters carried by a nearby stream or drainage ditch into the wetland, as wetlands often border these features. For this modification to be possible, the bottom of the adjacent ditch must be higher than the planned elevation of the water in the wetland at some point. The path of the ditch or small stream is changed by constructing a gradual turn uphill from the wetland so that waters flow downhill into the failed wetland. Because of the large amount of water that can flow into a wetland by this change, it may be necessary to widen and harden the spillway to protect the wetland from damage during storm events.

Gravels in the Bottom

Examining the bottom of a wetland can provide further clues as why it has not developed planned hydrology. Bring a shovel or a posthole digger to do some soil testing. The presence of scattered gravels and rock on the ground may show where someone was trying to build a wetland that would hold surface water from soils not suited for the purpose. Dig a couple of test holes in the area, within the zone that should contain deeper water and determine soil texture. If the shovel hits rock, then waters entering the wetland are soaking into the ground. Take a sample of soil about the size of a walnut, add a little water, and see how long a ribbon can be made between one's thumb and forefinger. If the soil feels gritty, and the ribbon breaks before it is two inches long, the soil may not have enough clay to hold water. Small areas of gravel or sand within large basins of clay will act like drains, much like holes in a leaky bucket. Should it be found that the bottom of the wetland is predominantly permeable soils, it may be possible to use a synthetic liner to complete the repair.

Bentonite

People have advocated using bentonite to seal leaky ponds for generations. Bentonite is simply finely-ground soil that is packaged and sold by the bag in feed stores. The product is used by first disking up the bottom of a leaky wetland, spreading bentonite over exposed soils, and then mixing with rakes or a disk. Instructions stress the need to keep the bentonite wet, so it can expand and seal in moisture.

This technique does not yield a permanent solution, however, because a thin layer of soil cannot form a lasting barrier over permeable ground. Compounding the problem is that bentonite generally has much lower clay content than the natural clays often found in the same area. Seldom can a thin ribbon be made from bentonite that is two inches long before breaking, while over four-inch-long ribbons can be made from native clays found in many places across North America.

Using bentonite is like trying to keep water out of a leaky basement with a thin layer of paint: water always finds a way through. One deer or elk wading in the water may puncture the bentonite, and a burrowing crayfish or muskrat can create huge holes in the thin layer. Fortunately, there are more reliable and, often, less expensive ways to repair a wetland.

The dozer works to lower and widen the spillway on this deep pond so it will function as an ephemeral wetland. Peninsulas were created from the soils that were removed to lower the dam.

Stream Restoration

Chapter 12

Restoring rivers and streams can provide great benefit to the environment by improving fish and wildlife habitat, and reducing flooding. Embarking on a stream restoration project involves careful planning and design, along with close attention to detail during construction.

Creeks and rivers were historically moved and straightened across most of North America. In most agricultural, urban, and forested areas it is not possible to find a creek or river that has not been changed. Rivers and streams altered years ago will continue to affect the environment for hundreds of years by:

- Forming vertical banks that wash and slide, causing significant erosion
- Deepening and widening, causing banks to collapse and wash downstream
- Supporting few and shallow pools that provide poor habitat for fish and wildlife
- Filling riffles with sediment and washing gravels downstream, exposing bedrock not suitable for fish spawning
- Washing fallen trees and boulders downstream, damaging roads and bridges
- Lowering the water table and draining historic wetlands
- Rapidly moving sediment and pollutants downstream

On landscapes where farming historically took place or are being actively worked now, expect creeks to have been moved. Major benefits were accrued to the farmer by moving streams with resulting enlargement and drying of fields. Looking for changes to creeks by comparing old with new photographs generally yields little information, as a majority of modifications were made prior to the 1930s.

Because there is a strong possibility that reference stream reaches, or sections that are used as examples

This natural braided stream contains leaves and regularly flows over its banks onto the floodplain.

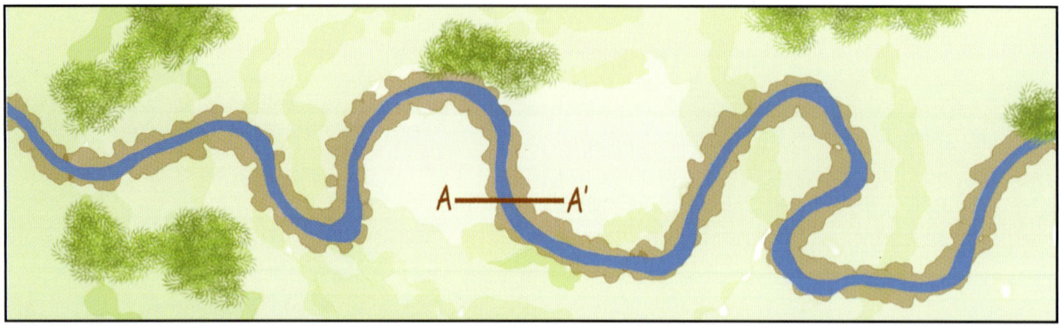

This natural creek is sinuous with meanders.

This natural creek is shallow where floodwaters readily flow across the floodplain.

This moved creek is straight with right-angle bends.

A straightened creek is deep with steep banks where floodwaters seldom flow across the floodplain.

Characteristics of natural and moved creeks

Characteristic	Natural Creek	Moved Creek
Map view	Sinuous	Straight with sharp turns
Location	Winds across valley	Follows base of hill
Channel	Braided with more than one channel on gradual slopes	One channel on gradual slopes
Vertical banks	Rare	Common
Substrate	Small gravels, including soil, and leaves	Bedrock, large rock and gravels
Wood in creek	Branches and small logs common	Infrequent large logs
Tributaries	Parallel main creek, entering at a shallow angle	Enter at right angles
Heavy rain event	Water flows across floodplain	Water stays in channel
Wetlands	Common on floodplain	Absent
Head-cuts	Rare	Common

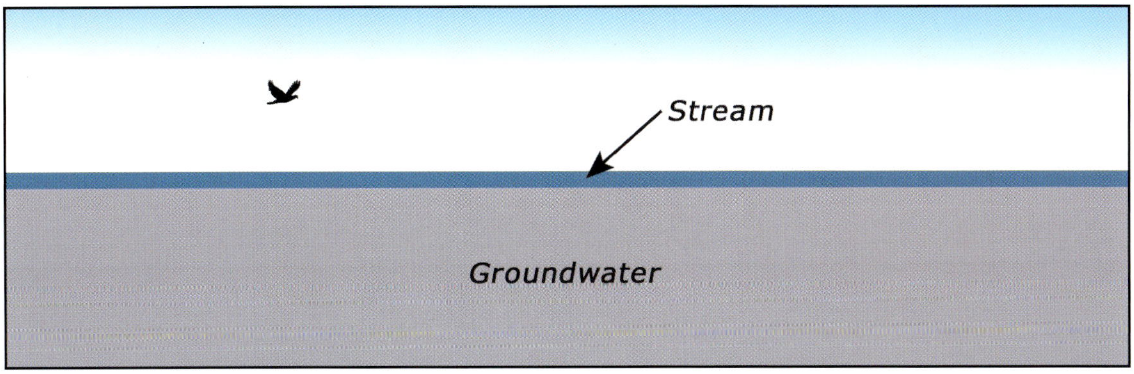

Groundwater elevation and the natural stream

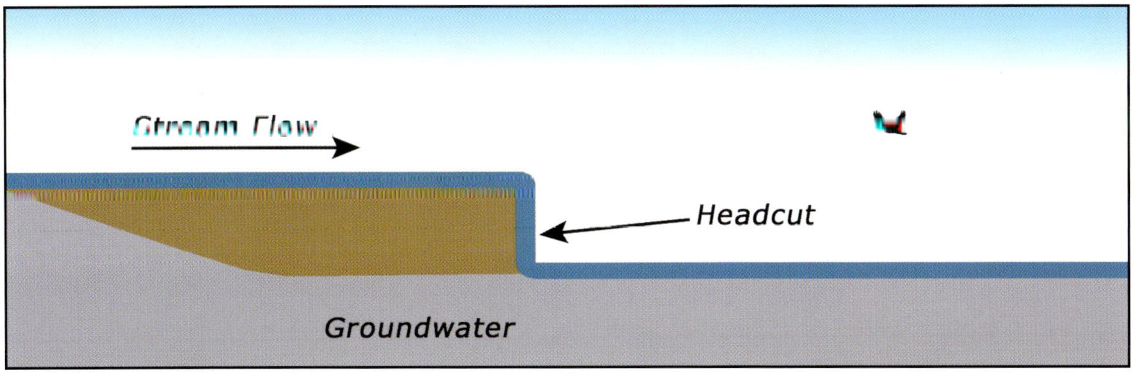

A head-cut will lower the elevation of groundwater as it moves upstream.

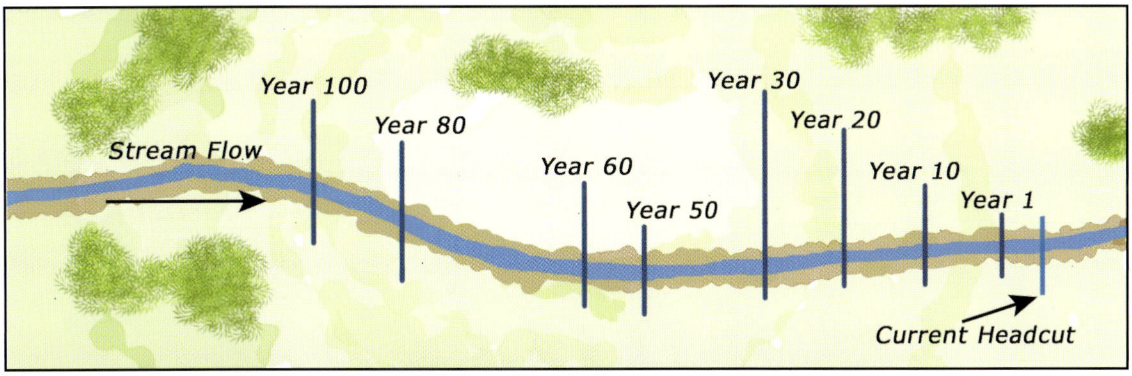

A head-cut forms when a downstream section of stream is straightened and continues to move upstream for many years.

The dog sits where water is flowing over a head-cut in a natural ephemeral stream. The head-cut began downhill after the lower part of the stream was moved, and has been advancing uphill about three feet each year.

were historically modified by road construction, farming, and logging operations, planned restoration actions should not necessarily attempt to duplicate characteristics of other possibly altered creeks and floodplains. Instead, projects should be designed to provide the structure and habitat that a diversity of fish and wildlife would use, with great attention being given to reducing the need for future maintenance.

Historic mill dams and logging dams greatly affected streams and wetlands on floodplains throughout the east. These dams trapped large quantities of sediment eroding from hillsides being farmed and logged without benefit of the best management practices. Wetlands and stream channels were buried by sediment 14 feet or more in some localities.

Clay drainage tiles are removed from a field on a six-percent slope prior to stream restoration near the Dix River in Lincoln County, Kentucky. Water begins filling the trench as the tiles are removed.

This is one of many small wetlands built near a newly-restored stream to provide fish habitat and water to the stream under low flow conditions.

Great care must be taken when removing dams in order to prevent massive amounts of erosion. When dams wash out, the affected stream will send a head-cut from the dam upstream through deposited sediments. The stream will cut through sediments down to its original elevation, washing huge amounts of soil downstream. The deep channels created by head-cuts are unstable and will widen over time, causing even more erosion.

Working with experts to design a stream restoration can be an exciting process involving a combination of science, mathematics, and art. Incorporating the following concepts can improve the quality and likelihood of success of stream restoration:

- Stream restorations should incorporate valley, floodplain, and channel sinuosity. Where creeks were historically channeled or moved, new, naturally-appearing meandering valleys and channels should be shaped across the valley.

- Structures should appear natural and be buried across the floodplain wherever possible. Avoid using structures that will not last.

- Wetlands of various depth and size should be built adjacent to streams, within and near the floodplain. These wetlands can be designed to capture and clean runoff, and supply waters to the creek under low flow conditions. Wetlands may include features that provide habitat for fish, aquatic organisms, amphibians, and reptiles. Forested, shrub, emergent, ephemeral, and wet-meadows should be included.

- Streams and wetlands should be designed for longevity, anticipating and eliminating the need for future maintenance.

- Where Cross-vanes and J-hooks are planned to control grade and improve habitat, daily construction monitoring is needed to limit future maintenance of these structures. Additional measures should be adopted to obtain grade control by using groundwater dams, burying rock structures across the floodplain at frequent intervals, and burying large woody debris and rock in the channel and floodplain.

- A goal of restoration should be to raise the water table and saturate the floodplain, providing conditions where a diversity of wetlands can be restored in the floodplain and valley. Sedges and rushes can be expected to colonize exposed soils on the floodplain that will control erosion better than sod and grasses.

- Groundwater dams should be used where soils are high in clay to block historic drainage structures, and to restore the elevation of the water table. These groundwater dams should be placed at intervals equal to every one-foot change in valley elevation. Groundwater dams can be used for grade control by imbedding large rock in their surfaces.

- Creeks should be designed so they will flow across the floodplain more often, except where they must be designed to transport sediment. Erosion can be controlled and stresses reduced by widening the floodplain and reducing the depth of stream channel. This can also improve habitat for fish and wildlife.

- Projects should be designed to reduce the landowner's concerns of flooding. Stream channels should be connected to their floodplain, and actions that minimize bank-full conditions are encouraged.

- Quarry rock that is used for grade control in the floodplain and creek channels should be buried and not be visible in the completed restoration.

- Look for opportunities to mine gravel and rock on-site, with borrow sites being reclaimed into wetlands when finished.

- Topsoil should be saved and spread over restored stream banks, wetlands, and the floodplain.

- An abundance of large and small woody debris should be added to restored streams, wetlands, and the riparian area, including the placement of vertical snags. Entire trees, including their root masses, should be used. Trees should be broken off, and not cut with a chainsaw. Where the Indiana bat is present, trees should not be felled during the roosting season.

- Action should be taken to control head-cuts in tributaries. This may involve the direct hardening of head-cuts, and the construction of splay areas and wetlands to serve as settling basins.

- Historic drainage practices should be identified, and steps taken to disable any continuing to function. Areas should be reshaped to eliminate ditches, "lands", dead furrows, and buried drainage structures consisting of wood, rock, clay, and plastic.

- Archeological sites must be protected from disturbance. They should be kept marked and must be avoided during construction.

- Only native species should be planted. Oaks adapted to wetlands should form the focus of the plantings.

Restoring streams can indirectly restore wetlands by creating conditions suitable for beaver occupation. The fast flowing waters found in moved streams can be difficult for beaver to dam. Waters in ditches often wash away beaver attempts to dam them. Restored streams are slower, with floodwater spreading across the floodplain. Beaver can successfully dam these streams, and in so doing, create a variety of wetlands.

Beaver often create a series of wetlands after a stream is returned to a more natural route, as they did along Meyers Fork Creek in Menifee County on the Daniel Boone National Forest. Meyers Fork and a tributary had been moved and straightened in the 1800s to create agricultural land. For years, the ditched tributary had been kept free of debris so that waters would not flood the planted field. The government purchased the land in the 1960s, and the fields grew up to sycamore and box elder trees.

Over time, the location where the tributary passed under a road through a culvert became blocked with branches and leaves, causing the creek to flow over an abandoned gravel road. The waters then spread over an old field in a braided pattern, eventually entering Meyers Fork 300 feet downstream from its original location. Beaver, quickly discovering and taking advantage of this natural opportunity, built a series of low dams across the braided flow between Meyers Fork and the road, creating stair-stepped forested and emergent wetlands.

Before the beaver took the initiative, the grown up fields had been examined for wetland restoration potential, but with soils high in sand and gravel, and the water table far below the surface, no suitable technique was identified. Fortunately, beaver blocking waters flowing across the valley created excellent habitat for wood ducks and spotted salamanders, and at no cost.

One of the main reasons this passive stream restoration was successful was that historic channels in the old field had not been completely filled. In most situations where creeks were moved to create fields, there is little evidence of historic stream channels, necessitating the reconstruction of valley, floodplain, and stream channels with restoration.

Digging Deep Test Holes

Digging deep test holes prior to construction can provide critical information that the designer can use to plan stream and wetland restoration projects. Soil survey maps generally do not provide the detailed information needed to plan these projects if they are located in valleys. An excavator large enough to dig test holes up to 25 feet deep should be used, as in most areas test holes must be dug deeper than a backhoe is able to reach. It is important that the person(s) designing the project be present when digging test holes to direct where they should be located, and to evaluate what is found. In one day, it should be possible to deep test a riparian area up to 30 acres in size, saving considerable expense, compared to financing a geo-technical exploration using core drilling machinery.

The benefits of digging deep test holes include:

- Ensuring a greater probability of project success
- Determining the elevation of groundwater; groundwater monitoring wells can be installed at the same time.
- Knowing the depth to bedrock for groundwater dam construction
- Measuring the depth and thickness of clay and gravel layers
- Identifying the size and cleanliness of gravel to see if it can be mined and used on-site
- Determining if groundwater or surface water construction techniques should be used
- Identifying filled wetlands
- Identifying buried organic layers
- Confirming the presence of buried drainage structures
- Identifying sites that can be used for groundwater recharge
- Determining how to raise the elevation of groundwater
- Helping to plan for the type and size of heavy equipment needed for construction
- Providing information to aid contractors in bidding so

they do not have to inflate costs to cover unknown conditions below the surface, which will, in all probability, lower construction costs

- Helping to reduce the number of contract change orders once construction begins

Steep, high, and eroding banks like this one are typically observed along streams that have been moved. Bedrock dominates the bottom of the stream, and waters flow over the banks only after the most severe floods.

This channeled stream was restored by creating a new sinuous floodplain. The banks along the stream are only six inches high, and flood waters will flow across the valley after a heavy rain.

This restored floodplain provides habitat to a diversity of aquatic plants. The logs and snags were placed as part of the stream restoration project.

Finding Funding

A large number of government agencies and nonprofit organizations fund wetland restoration projects across North America. Restoring wetlands and streams is a priority management practice for some agencies that manage natural resources. Because relatively few individuals plan wetland projects for reasons other than required mitigation, an individual's or group's interest in building a wetland to aid the environment should attract attention.

Before beginning a search for funds, there are some steps that can be taken to greatly increase one's prospect for success:

- Identify a location for building wetlands and mark it clearly on a map. Try to locate sites with minimum permitting requirements.
- List the main objectives for building the wetlands.
- Determine what technique can be used to build the wetlands.
- Obtain permission from the landowner to complete the project.
- Prepare a budget, showing partners, cash needs, and estimated in-kind contributions.
- Complete documentation for the project as required by the National Environmental Policy Act, and other rules and regulations.
- Summarize the above information in a two-page proposal which one can explain with clarity if asked to do so.

With a proposal in hand, one is well prepared to ask for financial help. The next step is to identify agencies and organizations that share an interest in completing the project. The local USDA NRCS office and state wildlife and fish office are good places to start.

Federal agencies that finance wetland projects include:

- U.S. Environmental Protection Agency (EPA)
- USDA Forest Service
- USDA Natural Resources Conservation Service (NRCS)
- USDI Bureau of Land Management (BLM)
- USDI Bureau of Reclamation
- USDI Fish and Wildlife Service
- National Oceanic and Atmospheric Administration (NOAA)

Some federal agencies provide grants for wetland and stream restoration. The announcements for all federal grants are available at: http://www.grants.gov

Most states have agencies that help finance wetland restoration projects. Those state agencies that manage natural resources are good places to start, including:

- Department of Natural Resources
- Conservation Districts
- Fish and Wildlife Management
- Water Quality
- Division of Forestry
- Mine Reclamation

There are a number of potential benefits to asking the government for help with one's wetland project. These can include free assistance with site selection and design, free aid in completing forms required to obtain necessary permits, and free supplies, such as seed, drainpipes, and water control structures. Agency personnel may be able to provide assistance in finding money to pay for the project, or better yet, actually provide the funding needed for a large portion of the wetland undertaking.

With all of these benefits of government assistance, one must ask why everyone does not ask government agencies for help. Some possible reasons are:

- One may have to build the wetland the agency's way and according to its schedule, losing some or all control of the project.
- The agency contact may not share one's enthusiasm and interest in the project.
- The project may be more expensive due to required design standards, workers compensation, and scale-wage requirements.
- One may not be able to choose the heavy equipment contractor to build the wetland because of open competition bidding requirements.
- It may be mandatory to sign a contract to maintain the site for a period of five to 30 years.
- The project may not be selected for funding after one has gone through all of the steps, and possibly some expense, to apply.

- One's family has a long-standing tradition of working the farm without government assistance.

Ducks Unlimited, Inc. is an international conservation organization whose mission is to conserve, restore, and manage wetlands and associated habitats for North America's waterfowl.[73] Ducks Unlimited will finance, design, and manage the construction of wetlands in the United States, Canada, and Mexico. The primary contact for discussing proposed wetland projects is the regional biologist for Ducks Unlimited in one's area.

In 2008, over 75,000 grant giving foundations in the US provided $45.6 billion dollars for worthy projects.[74] When adjusted for inflation, grant dollars have grown almost 80 percent since 1998. Each nonprofit foundation has clearly stated goals and objectives that guide its endowments. In many cases, wetland and stream restoration projects are a good match for charitable bodies with a mission that includes protecting the environment.

One can greatly increase the probability of receiving funding from a foundation by working with another nonprofit organization to request the funding. This is because a majority of foundations require that grant recipients also be nonprofit organizations. These organizations are to be exempt from federal income tax under section 501(a) of the Internal Revenue Code, and must meet the requirements described in section 501(c)(3).

Consider collaborating with a NRCS Resource Conservation and Development Program (RC&D) to complete the wetland project. RC&D's primary function is to plan and implement projects for natural resource conservation and community development. With goals including fish and wildlife conservation and environmental education, partnering with wetland and stream restoration projects can help the agency to fulfill its mission. Another bonus is that a RC&D can serve as a 501(c)(3) organization for projects it adopts.

Forming a new nonprofit organization is not that difficult, and has been done by a number of individuals and communities that were looking for financial help with wetland and watershed improvement projects. While completing the required IRS forms for establishing a nonprofit 501(c)(3) organization can take several days, the long term rewards are certainly worth the effort.

The best wetland project in the world cannot be completed unless someone finds the money required to reshape the soil. Fortunately, there is an enormous amount of funding available for wetland and stream restoration projects. By first preparing a proposal, and then, by working with agencies and nonprofit organizations, the vision of naturally-appearing and functioning wetlands can be obtained.

GLOSSARY

Backhoe: A machine mounted on rubber tires with a wide bucket on the front for carrying soil and a smaller bucket attached to an arm on the back for digging holes.

Compaction: Pushing soils together so tightly that there is little air between particles. Compaction is obtained by traveling over thin layers of soil that have the correct moisture content a number of times with heavy equipment. Soils are compacted to prevent water from soaking into the ground.

Construction level: A small telescope with a crosshair mounted on a stand used in combination with a survey rod to measure slope.

Construction fill: Soil, rock, wood, concrete, asphalt, and other waste materials used to pack wetlands and raise ground.

Core: The center of a dam that is made from soils high in clay, and compacted in layers.

Dam: A narrow ridge much like a roadbed that prevents water from flowing downhill, generally constructed from soils high in clay, and compacted in layers.

Dozer: A machine on tracks with a wide, large blade on the front designed for pushing and grading soils.

Emergent wetland: A wetland that has open water and aquatic plants, such as cattails and bulrushes, growing in and around it.

Elevation: The numerical reading observed on a survey rod when using a level; the change in height compared to a reference mark.

Ephemeral wetland: A wetland that dries annually or following a drought, the same as a vernal pond.

Excavator: A machine on tracks with a large bucket mounted on a mechanical arm that is designed for digging.

Forested wetland: A wetland that has trees growing in or around it.

Geo-textile: Thick fabric, similar to a blanket, used to protect synthetic liners from puncture when used for wetland construction.

Groundwater: Saturated soils that are below the surface of the ground.

Groundwater dam: A dam constructed below the above-ground dam to prevent water from leaving the wetland by flowing under the dam; the groundwater dam is used to raise the elevation of the water table for wetland and stream restoration projects. It is based on bedrock or an impermeable layer of clay, and built from clay compacted in layers.

Groundwater wetland: A wetland supplied primarily by water in the soil beneath the surface; a groundwater wetland exposes a high water table much like a hand-dug well does. Soils in the bottom of a groundwater wetland are not compacted, and may consist of gravel and sand.

Head-cut: A vertical drop-off in a stream or ditch that looks like a small waterfall. It is caused by the lowering of a stream or ditch downstream, or removal of gravel from the stream. A head-cut will form where the bottom of the stream channel was lowered suddenly, such as where a stream enters a straightened stream. Head-cuts begin downstream and move upstream over time, leading to the stream or ditch becoming deeper and wider, causing extensive erosion, lowering the water table on adjoining lands, draining adjacent wetlands, and filling wetlands located downstream.

Herps: A collective term used for describing reptiles and amphibians.

Lands: Parallel ridges of soil created to farm wetland; lands are formed by directionally plowing to raise ridges of soil above the elevation of water so the roots of planted crops or trees will not drown. Lands are separated by shallow ditches containing water, also known as dead furrows.

Laser level: A plastic box with batteries containing a rotating laser that is mounted on a stand; the laser level sends out a beam of light picked up by a receiver attached to a survey rod that is used for determining slope, requiring only one person for operation.

Level: Land that has little slope. Many call an area level if it has less than two percent slope.

Liner or **Synthetic liner:** Waterproof material made from EPDM or PVC that is aquatic-safe and used for building wetlands on coarse-textured soils where a high water table is not present.

Outlet: Where a buried drainage system made from clay tiles, plastic, wood, or rock empties into a ditch, stream, river, or lake; water flowing from the end of a pipe, combined with a short ditch and head-cut, often marks the location of an outlet. This shows that wetlands have been drained and streams placed underground in the surrounding landscape.

Riparian: Lands near a creek or river subject to flooding every 100 years, also known as the floodplain.

Soil auger: A hand-held steel rod attached to a large screw that can be used for taking soil samples at various depths below the surface.

Surface inlet: A direct pathway for water and air to enter a buried drainage system; a short, vertical section of pipe may be the only visible sign of a surface inlet. Surface inlets created from rock are not visible unless topsoil is removed. They are generally placed in the deepest part of a historic wetland, and at the upper reaches of a stream to collect waters before they can flow over a field.

Surface water wetland: A wetland constructed to contain rainfall and runoff by preventing water from soaking into the ground or flowing downhill; a surface water wetland contains water much like a bowl lined with a thick layer of compacted clay.

Synthetic liner: A large sheet of thick, strong waterproof material that is made from plastic or rubber.

Tile: A clay or cement pipe used to drain a wetland; generally 12 inches long, and from three to 18 inches in diameter, tiles were placed in dug ditches, end to end, and then covered with soil from two to 16 feet deep. Tiling eliminated standing water, and lowered the water table for crop production.

Tile probe: A long, small diameter steel rod with a steel tip mounted on a handle that can be used to locate buried drainage structures, the elevation of groundwater, presence of rock, and depth to bedrock.

Track-steer: A machine mounted on tracks with a large bucket on the front that is designed for scooping and moving soil. It is often called a Bobcat, which is one manufacturer of track-steers.

Vernal pond: A wetland that fills in the spring with snowmelt or from heavy rain and dries annually or following a drought.

Wetland: An area containing shallow water for all or part of the year, aquatic plants such as bulrushes and sedges that are adapted to living in water or saturated soil, and soils that may be gray, green, or black in color.

APPENDIX

SPECIFICATIONS FOR WETLAND CONTRACTS
SERVICE CONTRACT TYPE

Designed for contracts where the contractor is paid by the hour for each piece of heavy equipment that is provided with an operator for building wetlands. The landowners *Contract Representative* is expected to direct the construction of the wetlands under a service contract.

Scope of the Contract:

This contract requires a Dozer Rental and an Excavator Rental with Operator(s) for establishing wetlands in compliance with its terms, specifications, and provisions including furnishing labor, equipment, supervision, transportation, operating supplies, and incidentals.

Location and Description:

The project work area is located in ______________________________ County along Highway __________________________, near the community of __________________________________ in the state of ______________________________. The attached map shows the project location.

Evaluation Factors:

The *Contract Representative* will award a contract resulting from this solicitation to the responsible offeror whose offer conforming to the solicitation will be most advantageous to the landowner, price and other factors considered. In determining the responsibility of a quoter, the following criteria will be considered:

- The ability, capacity and skill of the Bidder to perform the contract or provide the services required.
- Whether the Bidder can perform the contract or provide the service promptly, or within the time specified, without delay or interference.
- The character, integrity, reliability, reputation, judgment, experience and efficiency of the Bidder.
- The quality of performance on previous contracts or services.
- The previous and existing compliance by the Bidder with laws and ordinances relating to the contract or service;
- The sufficiency of the financial resources and ability of the Bidder to perform the contract or provide the service.
- The quality, availability and adaptability of the goods or services to the particular use required.
- Whether the Bidder is in arrears for a debt or contract.
- The Bidder's overall price will be evaluated with past performance information for selection of an

offer that provides the best value to the *Contract Representative*.
- Such other information as may be secured by the *Contract Representative* having a bearing on the decision to award the contract.

Technical Specifications:

The Contractor shall furnish equipment and operator(s), and pay for all operating supplies and repairs. All equipment and personnel shall be located at the worksite and be ready for use as needed. The *Contract Representative* will determine when conditions are suitable for work. Work will not take place if soil conditions are not suitable for productive work. The equipment shall be the size and type needed for the establishment of wetlands and shall meet the minimum standards as follows:

Dozer:
105 HP minimum
20,000 lbs minimum

Excavator:
100- or 200- Series
80 HP minimum
22,500 lbs minimum
Bucket that is 42 inches wide or wider

The equipment shall be in good operating condition, cleaned of soils, and without hydraulic fluid, oil, or coolant leaks.

A tractor and trailer is required for moving equipment from one work site to another. The Contractor will not be paid to deliver equipment to the worksite when work begins, or to move equipment off the worksite when work ends. The *Contract Representative* will pay the Contractor to move between worksites at the hourly rate equal to the hourly rate paid for that piece of equipment being moved.

The person(s) operating heavy equipment on this contract shall be the same as those listed on the submitted **Operator Experience Questionnaire** form. Request for changes in person(s) operating equipment will be reviewed and approved in advance by the *Contract Representative* before any change in personnel is made by the Contractor.

The *Contract Representative* will determine when conditions are suitable for work and will monitor compliance with contract specifications. Each week the Contractor and the *Contract Representative* shall agree on a schedule of work for the next week. This schedule will be designed to make the most efficient use of time and equipment, based on predicted weather and site conditions. The Contractor may work up to 10 hours per day, Monday-Saturday.

The *Contract Representative* will seed and mulch exposed soils where work has been completed.

Administration:

This contract will be administered by the *Contract Representative*. Prior to commencement of work, the *Contract Representative* will arrange a meeting with the Contractor to discuss the contract terms and work performance requirements. A work progress schedule will be developed at this meeting.

Before work starts, preferably at the pre-work conference, the Contractor shall designate in writing an individual to act for him in his absence. The Contractor or the Designated Representative shall be on the job whenever work is in progress.

In the absence of the Contractor, the Designated Representative will take action as required to keep the job in progress under the terms of the contract. The presence of the Contractor's representative on the job will not relieve the Contractor of responsibility for satisfactory performance.

Inspection and Acceptance:

Inspection will be made as work progresses to ensure compliance with contract terms, conditions and specifications. Any work not being properly performed will be brought to the immediate attention of the landowner by the *Contract Representative*. The Contractor shall take prompt action for any deficiencies or deviations that may occur under this contract. The *Contract Representative* will promptly advise the Contractor of his acceptance or rejection of work performed. Unacceptable work must be corrected by the Contractor before payment is made.

Labor Laws:

The Service Contract Act of 1965 as amended applies to this procurement and Wage Determination No. 80-0151 (Rev 31) is attached.

Equipment Furnished by the Contractor:

The *Contract Representative* is not liable for loss, damage, or destruction of equipment furnished under this contract, except for such loss, damage, or destruction resulting from the negligent or wrongful act(s) of *Contract Representative* employee(s) while acting within the scope of their employment. All claims arising under this clause will be submitted to the landowner.

Repair, Maintenance, Supplies, and Operation:

The Contractor shall be responsible for all repairs, maintenance, and operating supplies that are required in the use of his equipment. Should the need for repairs develop, such repairs shall be accomplished at no additional cost to the *Contract Representative*. The *Contract Representative* is not responsible for down-time on equipment that is not in service due to repair, maintenance or no requirement of its services at that moment. Each day both the *Contract Representative* and Contractor, or contractor's representative, will sign the equipment log book recording the daily hours that each piece of equipment is in operation.

Required Work Dates:

Work will begin on ____________________ and be completed on or before ____________________.

Award Determination:

The *Contract Representative* will award a contract resulting from this solicitation to the responsible offeror whose offer conforming to the solicitation will be most advantageous to the landowner, price and other factors considered.

Evaluation:

The following factors will also be used to evaluate offers:

- Whether the heavy equipment requested meets specifications
- Experience building wetlands with the dozer and/or excavator
- past performance operating the dozer or excavator
- Price quoted per hour

Experience:

Each offeror shall provide a written summary of the most recent experience (past 3 years) in similar efforts. An **Experience Questionnaire** is attached for the offeror's convenience in submitting a quote. Please include this questionnaire and equipment Information form with the quote to the address listed.

Past Performance:

Past performance is measured by the offeror's compliance with federal, state and local laws and regulations, and customer satisfaction. The offeror shall provide references for the most recent contracts relevant to the type of work being solicited which support past performance. An **Experience Questionnaire** is attached for your convenience in submitting your quote. Please submit this questionnaire to the address listed.

Equipment Requirements:

The equipment shall be the size and type needed for the job. An **Equipment Information Form** is attached for your convenience in submitting your quote. Please include this form with your quote to the address listed.

Note:

There will be no formal site showing for this project. Quoter's are encouraged to utilize the information provided in this solicitation prior to submitting a quote. You may contact ______________________ at ___________________________ for technical questions, and for additional information.

EQUIPMENT INFORMATION FORM

Submit with your quote to the following address:

____
____
____
____
____

Company Name:____

Address:____

Phone:____________________ Cell Phone: ____________________

Fax: ____________________ E-Mail: ____________________

Federal ID # or SS# ____

Duns #____

Dozer

Make: ____________________

Model: ____________________

Year: ____________________

Weight: ____________________

Horsepower: ____________________

Price per hour: $____________________

Excavator

Make: ____________________

Model: ____________________

Year: ____________________

Weight: ____________________

Horsepower: ____________________

Price per hour: $____________________

Please note that the horsepower and weight listed by the manufacture for the model and year listed will preside.

EXPERIENCE QUESTIONNAIRE

Submit with your quote to the following address for each operator:

____
____
____
____
____

Business submitting bid: ____________________________ Phone: ____________________

Name of individual operating equipment: ______________________________________

Please answer the following questions about this individual:
1) Years experience operating dozer: _______________
2) Years experience operating excavator: _________________

3) Please describe this person's experience building wetlands:

3) Please list the names and phone numbers of three individuals we may contact about this operator and his or her capabilities:

A) Name: __________________ Phone: __________________ 2nd Phone: _______________

B) Name: __________________ Phone: __________________ 2nd Phone: _______________

C) Name: __________________ Phone: __________________ 2nd Phone: _______________

SPECIFICATIONS FOR WETLAND CONTRACTS
CONSTRUCTION CONTRACT TYPE

Under a construction contract the Contractor is paid by the job for the wetlands that are built according to written specifications and drawings. The Contractor is expected to direct the construction of the wetlands.

SURFACE WATER TECHNIQUE

Description of Work:

Shallow water areas up to 18 inches deep will be constructed at locations shown by the attached map (s) and as marked on the ground. The wetlands will vary, and be round, oval, or crescent in shape. An excavator (100-Series, at least 80 HP and 22,500 lbs) and dozer (at least 105 HP and 20,000 lbs) working together are required for this job.

Contractor will furnish equipment and operators. Contractor will pay for all operating supplies and repairs. The equipment shall be in good operating condition, cleaned of soils, and without hydraulic fluid, oil, or coolant leaks. Work will begin on or before ______________________________ and will be completed by __________________________.

Specifications:

The contractor will complete the work in a manner that reduces erosion. Straw bales will be placed across ditches, and depressions and diversion ditches will be dug, as needed, to contain sediment. Only one worksite may be opened and operated at any one time without previous approval from the *Contract Representative*. Wet sites may be ditched and drained prior to construction.

Remove trees and shrubs from designated pool area of the wetland, and save adjacent to area for placement in wetland. Remove topsoil from designated area and save adjacent to area for placement in wetland. The *Contract Representative* will determine when topsoil has been removed to a sufficient depth.

Excavate a trench on the location of the dam as designated by the *Contract Representative*. The trench will be dug on the location of the dam, regardless of the height of the dam. The trench will approximately follow the lower two-thirds perimeter of the designated pool area for the wetland. The trench will be at least 12 feet wide, and extend down to intercept buried drain lines and reach the bottom of crayfish holes. The *Contract Representative* must be present when this step is being completed. The Contractor can expect to dig the trench to an average depth of __ feet.

Soil removed from the trench that is suitable clay for replacement in the trench to form the groundwater dam will be deposited along the edge of the trench, inside the future wetland. Soil removed from the trench that is not suitable (gravel or sand) for placement in the trench will be piled along the edge of the trench, outside the wetland. The *Contract Representative* will make the determination of which soil is suitable for use in the groundwater dam and dam.

The trench will be filled and compacted, as it is being dug, with clay soils to reduce cave-ins. Clay soils that are suitable for the groundwater dam and dam may be taken from inside the designated pool area of the wetland. Soil will be removed beginning at the back edge of the stripped area and proceeding over the designated pool area to maintain gradual slopes, and to maximize the surface area of water in the wetland. The back slope of the dam may be constructed of non-clay soils removed from the trench.

The soil in the groundwater dam and dam will be packed by running over each layer with a dozer or excavator four or more times as determined by the *Contract Representative* until suitable compaction is obtained. Soil placed in the trench and dam will be compacted in layers no greater than six inches thick. Soils that are wet may be mixed with dry clays, removed and replaced, ditched and drained, or air-dried to achieve compaction.

The dam will be built to an elevation to obtain a maximum depth of water of 18 inches in each wetland, with depths ranging from one to 18 inches. The top of the dam will be at least 12 feet wide and be made level (± 0.1 feet). The inside and backside of the dam will have slopes of five or less percent.

Sides of the designated pool area will be sloped so that they vary from one to five percent.

A spillway will be constructed on each wetland with an entrance elevation that is 0.5 feet below the top of the dam. The spillway will be located at one end of the dam, so that water will flow from the wetland and away from the dam, down a gradual slope over undisturbed soils. The location, slope, and shape of the spillway will be designated by the *Contract Representative*.

Spread the topsoil that has been saved over slopes and around excavation, but do not compact. The bottom of the wetland will not be graded, but left rough with humps and dips.

Small islands will be formed within constructed wetlands. The number of islands will vary from one to 20 per wetland. The islands will vary in size from ten to 30 feet across, and be round, oval, or crescent in shape. The top of each island will extend one foot or more above the elevation of water in the wetland being constructed. Slopes surrounding each island will be no steeper than five percent, and soils will not be compacted to facilitate tree and shrub growth.

Spread shrubs and trees that were saved within and around the wetland, anchoring portions with soil so they do not wash away. Each wetland should contain a minimum of two large trees, live or dead, that are laid horizontally. The trees will be 9 inches or larger in diameter at breast height, with root wads and branches attached. The trees will be placed in a random, naturally-appearing manner so that some are entirely in waters of the wetland, while others are only partially in the water. Trees and portions of trees placed in wetlands will not be cut with a chainsaw, but rather broken off to appear natural.

GROUNDWATER TECHNIQUE

Description of Work:

Shallow areas of water will be constructed that will be supplied primarily with groundwater at locations shown by the attached map(s) and as marked on the ground. Excavations will be made from three to six feet deep, with gradual slopes, and no obvious dam. The finished area of water in the wetlands will average one-half the size of the clearing limits, which will be marked by colored flags. Groundwater elevation on the worksites ranges from six to 24 inches below the surface. The excavations will be round, oval, or crescent in shape. An excavator (100-Series, at least 80 HP and 22,500 lbs) and a dozer (at least 105 HP and 20,000 lbs) working together are required for the job.

Contractor will furnish equipment and operators. Contractor will pay for all operating supplies and repairs. The equipment shall be in good operating condition, cleaned of soils, and without hydraulic fluid, oil, or coolant leaks. Work will begin on or before ______________________________ and be completed by ______________________________.

Specifications:

The Contractor will complete the work in a manner that reduces erosion. Straw bales will be placed across ditches, and depressions and diversion ditches will be dug, as needed, to contain sediment. Only one worksite may be opened and operated at any one time without previous approval from the *Contract Representative*. Wet sites may be ditched and drained prior to construction.

Remove trees and shrubs from the designated area, and save in an adjacent location for replacement in the wetland. Remove topsoil from the designated area, and save in a nearby space for replacement in the wetland.

A narrow trench, one bucket wide and eight feet deep will be dug along the lower two-thirds perimeter of the designated area to identify possible buried drainage structures. The *Contract Representative* must be present when this step is being done. If no drainage structures are found, the ditch will be filled and compacted using the bucket with the soils removed. Where drainage structures are found, they will be removed and blocked for a distance of 20 feet within or outside of the designated pool area, so that they cannot drain the wetland being built. The ditch surrounding the exposed 20-foot length of drainage structure will be widened to 12 feet. The widened ditch will be filled with clays compacted in layers no greater than 6 inches thick at 95 percent standard proctor density. Soils may be compacted by running over each layer with the dozer or excavator, or by using the bucket of the excavator. Clays that are wet may be mixed with dry clays, removed and replaced, ditched and drained, or air-dried to achieve compaction.

Excavate the planned wetland to the required elevation. The center of the excavation will extend down into groundwater, which is generally indicated by sand and silts. Soil removed in creating the excavation will be spread over the lower portion of the designated area to form a dam with a level top, 12 feet or wider, with gradual back slopes that end at the lower edge of the designated area. The elevation of the dam will be no greater than the upper edge of the designated area, to avoid creating a rim of soil around the wetland, so that runoff can enter the wetland. Soils removed in creating the excavation may also be added to existing slopes around the wetland, and be used to fill low areas to improve the appearance and size of the wetland.

The excavation will initially be shaped like a large satellite dish. Sides of the excavation will be sloped to average five percent. In addition, where the wetland will be connected to flowing water, such as a channel, stream, or swale, areas up to 1,000 square feet will be created with a one percent slope. These more level areas will be located along the slopes leading into and out from the wetland to provide shorebird habitat.

Avoid compacting soils within or surrounding the excavation. The bucket on the excavator will be used to finish the digging to avoid packing the bottom. Tracked or wheeled equipment is not allowed in the completed excavation to reduce the possibility of compacting soils, or of mixing clays in the depression, Should soils become compacted in the excavation, they will be loosened by ripping to a depth of two feet with an excavator bucket or a sub-soil ripper.

Spread topsoil that has been saved over slopes and around excavation, but do not compact. The bottom of the wetland will not be graded, but will be left rough with humps and dips.

Small islands will be formed within constructed wetlands. The number of islands will vary from one to 20 per wetland. The islands will vary in size from ten to 30 feet across, and be round, oval, or crescent in shape. The top of each island will extend one foot or more above the elevation of water in the wetland being constructed. Slopes surrounding each island will be no steeper than five percent, and soils will not be compacted to facilitate tree and shrub growth.

Spread shrubs and trees that were saved within and around the wetland, anchoring portions with soil so they do not wash away. Each wetland should contain a minimum of two large trees, live or dead, that are laid horizontal. The trees will be nine inches or larger in diameter at breast height, with root wads and branches attached. The trees will be placed in a natural, random manner so that some are entirely in waters of the wetland, while others are only partially in the water. Trees and portions of trees placed in wetlands will not be cut with a chainsaw, but rather broken off to appear natural.

WETLAND RENOVATION AND REPAIR

Description of Work:

Work involves rebuilding, reshaping, and constructing groundwater dams for repairing shallow water wetlands so that they will hold water as planned, according to locations shown on the attached map (s). Worksites are constructed wetlands that have been mowed on a regular basis, or have grown up to small-diameter trees. An excavator (100-Series, at least 80 HP and 22,500 lbs) and a dozer (at least 105 HP and 20,000 lbs) working together are required for the job.

Contractor will furnish equipment and operators. Contractor will pay for all operating supplies and repairs. The equipment shall be in good operating condition, cleaned of soils, and without hydraulic fluid, oil, or coolant leaks. Work will begin on or before ______________________________ and be completed by ______________________________.

Specifications:

The contractor will complete the work in a manner that reduces erosion. Straw bales will be placed across ditches, and depressions and diversion ditches will be dug, as needed, to contain sediment. Only one worksite may be opened and operated at any one time without previous approval from the *Contract Representative*. Wet sites may be ditched and drained prior to construction.

Remove small trees and shrubs in a wide strip behind each dam to provide room for waste material, and to shape a more gradual slope on the back of the dam. This action will generally involve clearing shrubs and small trees that have grown up near the dam since it was first constructed. In some cases, the location of the dam will be moved from its current location, and a limited number of larger trees will be removed for its construction. Areas where larger trees are to be cleared will be designated by the *Contract Representative*, most often by hanging flagging from vegetation. Larger trees that are removed will be piled in locations adjacent to the worksite for later placement in each completed wetland.

Remove vegetation and topsoil from the top of the existing dam, the inside slope of the dam, and a strip approximately 100 feet wide in front of the dam. This material may be pushed behind the dam and used to place a more gradual slope on the backside of the dam. The *Contract Representative* will determine when topsoil has been removed to a sufficient depth.

Remove vegetation and topsoil from the bottom of each wetland and windrow on higher ground along the upper edge of each wetland, as agreed upon with the *Contract Representative*. This material will be spread in the bottom of each wetland after it is rebuilt.

Prepare a level bench for the excavator to dig by removing the inside toe of the dam. Providing the soil is high in clay, it may be piled along the inside edge of the trench that is being dug for later placement in the groundwater dam. This action will create a level path for the excavator to follow for digging the trench beneath the inside toe of the dam. The excavator will be working next to a high wall whose edge is the top of the dam.

Use the dozer and the excavator to create a deep trench on the constructed bench. The location and length of the trench will be designated by the *Contract Representative* by the use of colored flags. The trench will be dug for the entire length of the dam, even where the dam is only one inch high. The trench starts on high ground where the dam begins, continues the entire distance of the dam, and ends at the same elevation where the dam again ties into higher ground. The trench will be constructed to a minimum width of the dozer blade, being approximately 12 feet wide, and extend down to bedrock or to an impermeable layer of soil as determined by the *Contract Representative*.

Soil removed from the trench that is suitable for placement in the groundwater dam will be deposited inside the wetland being rebuilt. The soil in the trench that is not suitable for the groundwater dam will be removed and placed on the backside of the dam, not inside the wetland. This unsuitable material will be used to place a more gradual slope on the backside of the dam. The *Contract Representative* will make the determination of which soil is suitable for use in the groundwater dam and dam. The groundwater dam, and the dam will only be constructed when the *Contract Representative* is on-site.

Existing drainpipes will be removed and replaced following construction of the groundwater dam.

Soil that is high in clay will be placed back in the trench created for the groundwater dam, and packed with the dozer in layers no greater than six inches thick. The soil in the groundwater dam and the dam will be packed by running over each layer with a dozer or an excavator four or more times, as determined by the *Contract Representative,* until suitable compaction is obtained.

Soil for the groundwater dam and dam will be taken from inside the designated pool area of each wetland, from the section previously stripped of vegetation and topsoil. Soil for the groundwater dam and the dam will be removed starting from the back edge of the cleared area and proceeding over the future pool area to maintain gradual slopes, and to maximize the surface area of water in the new wetland. Should a wall of the trench collapse and place gravel in the groundwater dam, the gravel must be removed and be replaced with suitable material. It is often necessary for the dozer to immediately push soil into the cleaned trench as soon as it is removed by the excavator to prevent cave-ins from occurring. This work must be closely coordinated between the dozer and excavator operators to be successful.

The dam will be reshaped to the following specifications. The top of the dam will be the width of one dozer blade, approximately 12 feet or wider. The inside slope of the dam will be a five or more percent gradual; the backside of the dam will be five or more percent gradual. The dam will be built to the elevation of the existing dam, or lower as designated by a benchmark set by the *Contract Representative*, generally a nail placed in a tree or the top of a wooden stake. The top of the dam will be made level within +/- 0.1 feet of the elevation of the benchmark.

Where space is limited along the back of the dam, permeable material removed from constructing the groundwater dam may be spread over the top of the finished dam and within the wetland.

The contractor will install a Schedule 35 SDR PVC gasketed drainpipe in the wetland. The drainpipe will be placed so that it removes water from the completed wetland. Soils will be packed around the drainpipe to prevent seepage. The bottom of the wetland will be shaped so that the drainpipe can remove waters, much like a bathtub drain. A water control structure will be attached to the drainpipe, with assistance from the *Contract Representative*. The drainpipe and water control structure will be provided and delivered on-site by the *Contract Representative* at no charge to the contractor.

A spillway will be constructed on the wetland to an elevation of six inches below the top of the dam, at one end of the dam, so that water will flow out of the wetland over undisturbed soils. The location, slope, and shape of the spillway will be designated by the *Contract Representative*.

The saved topsoil and vegetation will be spread over the bottom of the completed wetland. The large trees that were saved will be placed in the bottom of the wetland at locations designated by the *Contract Representative*.

ROADBED WETLANDS

Description of Work:

Work involves building shallow water wetlands in roads that will trap sediment, contain water, and provide habitat for fish or wildlife species. These wetlands will be located in roads where water is flowing down the road, where water is standing in the road, and/or where ruts are located. Clusters of wetlands are also acceptable.

Contractor will furnish equipment and operators. Contractor will pay for all operating supplies and repairs. The equipment shall be in good operating condition, cleaned of soils, and without hydraulic fluid, oil, or coolant leaks. Work will begin on or before ______________________________ and be completed by ______________________________.

Specifications:

Wetlands can be from 10 to 100 feet long and from 10 to 36 inches deep.

The sides of the wetland will be sloped. Gradual slopes will be placed along the uphill edge of the wetland.

A spillway will be constructed for each wetland that prevents water from flowing down the road, and that prevents erosion to the wetland.

Gravel will be scraped from each worksite prior to construction.

Topsoil will be removed and saved for later spreading within and around each wetland depression.

A hole that is approximately four feet deep will be dug across the area of the planned wetland, unless bedrock is encountered that is less than four feet below the surface. The hole will cut through buried wood, roots, topsoil, pipe, and gravel layers, much like the core on a larger wetland.

Where clay soils are encountered, the clay soils will be placed back in the depression in layers no thicker than six inches, with each layer being compacted by heavy equipment. These wetlands will be supplied primarily by surface water runoff.

Where gravel or sandy soils are encountered, soils will be placed back into the depression to the desired depth. The soils will not be compacted, as these wetlands will be supplied primarily by groundwater, such as from springs.

Excess soils and gravel may be spread downhill from the wetland depression, generally over the roadbed. These soils will not be compacted so they will grow vegetation. Avoid creating dams from these soils.

Topsoil that was saved will be spread within and around the wetland depression.

Logs and branches will be placed in and around the completed wetland depression.

REFERENCES

1. Kathryn Flynn, *Understanding Wetlands and Endangered Species: Definitions and Relationships,* Alabama Cooperative Extension System, ANR-979, 4, at http://www.aces.edu/pubs/does/A/ANR-0979/ANR-0979.pdf (retrieved on November 27, 2006).
2. Thomas E. Dahl, *Wetlands Losses in the United States, 1780's to 1980's* (Washington, D.C.: U.S. Department of the Interior, Fish and Wildlife Service, 1990), 1.
3. Ibid., 6.
4. Thomas E. Dahl, *Status and Trends of Wetlands in the Conterminous United States 1998 to 2004*, (Washington, D.C: U.S. Department of Interior; Fish and Wildlife Service, 2006), 44.
5. Thomas R. Biebighauser, *Wetland Drainage, Restoration, and Repair*, (Lexington, Ky.: University Press of Kentucky, 2007), 36-38.
6. Dahl, *Status and Trends of Wetlands in the Conterminous United States 1998 to 2004*, 66, 72.
7. Lewis M. Cowardin, V. Carter, F.C. Golet, and E.T. LaRoe, *Classification of Wetlands and Deepwater Habitats of the United States* (Washington, D.C.: U.S. Department of the Interior, Office of Biological Services, Fish and Wildlife Service, 1979), 103 pages.
8. Henry F. French, *Farm Drainage* (New York, NY: Orange Judd Company, 1903): 320-321.
9. Robert B. Cromer, Joeseph D. Lanham, Hugh H. Hanlin, *Herpetofaunal Response to Gap and Skidder-Rut Wetland Creation in a Southern Bottomland Hardwood Forest*, Forest Science 48(2), 2002, 412.
10. Manly Miles, *Land Draining, A Handbook for Farmers on the Principles and Practice of Farm Draining* (New York: Orange Judd Company, 1893), 184.
11. French, *Farm Drainage*, 316.
12. French, *Farm Drainage*, XI.
13. B.J. Allred, N.R. Fausey, L.Peters, Jr., C.Chen, J.J. Daniels, and A.Yaun, *Detection of Buried Agricultural Drainage Pipe with Geophysical Methods*, Applied Engineering in Agriculture (2004) 20(3): 307-318.
14. John H. Klippart, *The Principles and Practice of Land Drainage* (Cincinnati: Robert Clarke, 1861), 238.
15. Biebighauser, Wetland Drainage, Restoration, and Repair, 46, 55-57, 93-94.
16. Guenter A. Schuster, interview by author, Richmond, Ky., January 16, 2006.
17. Guenter A. Schuster, interview by author, Richmond, Ky., January 16, 2006.
18. Guenter A. Schuster, interview by author, Richmond, Ky., January 16, 2006.
19. John Johnstone, *An Account of the Mode of Draining Land*, 3rd edition (London: Richard Phillips, 1808), 86-87.
20. Klippart, *The Principles and Practice of Land Drainage*, 249-250; Elliott, *Engineering for Land Drainage, A Manual for the Reclamation of Lands Injured by Water*, 3rd edition (London: Wiley, 1919), 326-327.
21. J.L. Richardson and M.J. Vepraskas, *Wetland Soils*, (Boca Raton, Fla.:CRC Press, 2001), 173.
22. Ibid., 169.
23. Thomas E. Dahl, *Wetlands losses in the United States 1780's to 1980's*, 8-9.
24. Biebighauser, *Wetland Drainage, Restoration, and Repair*, 98.
25. *Natural Resources Inventory*, 2002 Annual NRI, USDA 2004, at http://www.nrcs.usda.gov/technical/NRI/ (retrieved February 20, 2005).
26. Merlin Spencer, interview by author, Frenchburg, Ky., September 11, 2007.
27. Alice L. Thompson and Charles S. Luthin, *Wetland Restoration Handbook for Wisconsin Landowners, 2nd Edition* (Madison, WI: Bureau of Integrated Science Services, Wisconsin Department of Natural Resources, 2004), 41.
28. David Scott, interview and e-mail correspondence with author, Sooke, BC, July 21, 2005
29. Biebighauser, *Wetland Drainage, Restoration, and Repair*, 199-216.
30. Karen L. Ripple and Edgar W. Garbish, *POW! The Planning of Wetlands, An Educator's Guide*, (Environmental Concern Inc., St. Michaels, Maryland, 2000), 31-45.
31. Christopher A. Taylor and Guenter A. Schuster, *The Crayfishes of Kentucky*, Illinois Natural History Survey Special Publication Number 28, 2004, 13-15.

32. Thomas R. Biebighauser, Density of crayfish burrows measured in a 70-acre field near the Dix River in Lincoln County, Ky., January 9, 2008.
33. Christopher A. Taylor and Guenter A. Schuster, *The Crayfishes of Kentucky*, Illinois Natural History Survey Special Publication Number 28, 2004, 15.
34. Matthew W. Hyatt, *Investigation of Crayfish Control Technology*, Final Report, Cooperative Agreement No. 1448-20181-02-J850, Arizona Game and Fish Department, Wildlife Management Division/Research Branch, 2221 West Greenway Road, Phoenix, AZ 85023, 1.
35. S.M. Blomquist, *Control of an Introduced Crayfish, Orconectes virilis, with Traps and Dipnets*, Nongame and Endangered Wildlife Program Technical Report 216, Arizona Game and Fish Department, Phoenix, Arizona, 2003.
36. Guenter A. Schuster, interview by author, Richmond, Ky., January 16, 2006.
37. Buhlmann, Kurt, Tracey Tuberville, and Whit Gibbons. *Turtles of the Southeast*. The University of Georgia Press, 2008, 13.
38. Reed, Sherwood C., Ronald W. Crites, E.Joe Middlebrooks. 1995. *Natural Systems for Waste Management and Treatment*, Second Edition. McGraw-Hill, Inc., 29.
39. Nolte, B.H. and N.R. Fausey. *Soil Compaction and Drainage*. The Ohio State University, Publication AEX-301, at http://ohioline.osu.edu/b301/index.html (retrieved February 3, 2006), 1.
40. Ray Frolek, Frolek Cattle Company, interview by author, Barnhartvale, British Columbia, October 1, 2006.
41. Mathiak, Harold A. 1965. *Pothole Blasting for Wildlife*. Publication 352, Wisconsin Conservation Department, Madison, Wisconsin, p. 4.
42. Mathiak, Harold A. 1965. *Pothole Blasting for Wildlife*. Publication 352, Wisconsin Conservation Department, Madison, Wisconsin, p. 16.
43. Elizabeth Byers, field interview by author, Elkins, Wv., April 8, 2010.
44. Shane Jones, field interview and e-mail correspondence with author, Elkins, Wv., April 8 and 14, 2010.
45. Elizabeth Byers, field interview and e-mail correspondence with author, Elkins, Wv., April 8 and 12, 2010.
46. Elizabeth Byers, field interview by author, Elkins, Wv., April 8, 2010.
47. Elizabeth Byers, field interview and e-mail correspondence with author, Elkins, Wv., April 8 and 12, 2010.
48. Elizabeth Byers, field interview by author, Elkins, WV., April 8, 2010.
49. Brodman, Robert, Jill Ogger, Megan Kolaczyk, Rachel A. Pulver, Amanda J. Long, and Tracey Bogard. *Mosquito Control by Pond-Breeding Salamander Larvae*. 2003. Herpetological Review 34(2), 2003, pp. 116-119.
50. Todd Pester, *Allelopathic Effects of Rye (Secale Cereale L.) And Their Implications for Weed Management - A Review*, Colorado State University, at: http://www.colostate.edu/Depts/Entomology/courses/en570/papers_1998/pester.htm (retrieved: February 14, 2010).
51. April D. Haight, *Evaluation of the Success of Constructed Wetlands in the Cave Run Lake Watershed*, Master of Science thesis, Morehead State University, Morehead, Ky., 1996.
52. *Common rush Juncus effuses Plant Fact Sheet*. USDA Natural Resources Conservation Service, at:
53. Mike Dorcas and Whit Gibbons, *Frogs and Toads of the Southeast,* (The University of Georgia Press, 2008), 185.
54. Elke Wind, field interview by author, Portland, Or., February 15, 2009.
55. Elke Wind, field interview by author, Portland, Or., February 15, 2009.
56. John MacGregor and James Kiser, field interview by author, Beattyville, Kentucky, January 23, 2009.
57. Austin Jenkins, telephone conversation with author, Morehead, Ky., February 15, 2008.
58. R.S. Butler, R.J. DiStefano, and G.A. Schuster, *Crayfish: an overlooked fauna.* Endangered Species Bulletin 28:10-11, 2003.
59. Horton H. Hobbs, Jr., and Mike Whiteman, *Notes on the burrows, behavior, and color of the crayfish Fallicambarus (F.) devastator (Decapoda: Cambaridae)*, The Southwestern Naturalist, Volume 36, Number 1, March, 1991: 128.
60. John MacGregor, field interview by author, Beattyville, Kentucky, January 17, 2006.
61. Guenter A. Schuster, interview by author, Richmond, Ky., January 16, 2006.
62. Matthew W. Hyatt, *Investigation of Crayfish Control Technology*, Final Report, Cooperative Agreement No. 1448-20181-02-J850, (Arizona Game and Fish Department, Wildlife Management Division/Research Branch, 2221 West Greenway Road, Phoenix, Az., 85023), 25.
63. Guenter A. Schuster, interview by author, Richmond, Ky., January 16, 2006.
64. Environmental Laboratory, *Corps of Engineers Wetlands Delineation Manual*, Technical Report Y-87-1, (U.S. Army Engineer Waterways Experiment Station, Vicksburg, Mississippi, 1987).
65. Nationwide Permit (27) Stream and Wetland Restoration Activities (3/18/2002), at http://www.nao.usace.army.mil/technical%20services/Regulatory%20branch/NW_enclosures/NW-27.pdf (retrieved on March 26, 2011).
66. Nationwide Permit (27) Stream and Wetland Restoration Activities (3/18/2002), at http://www.nao.usace.army.mil/technical%20services/Regulatory%20branch/NW_enclosures/NW-27.pdf (retrieved on March 26, 2011).

67. Nationwide Permit (27) Stream and Wetland Restoration Activities (3/18/2002), at http://www.nao.usace.army.mil/technical%20services/Regulatory%20branch/NW_enclosures/NW-27.pdf (retrieved on March 26, 2011).
68. Clean Water Act Sections 401 and 404, at http://www.nmenv.state.nm.us/SWQB/33cfr_320.pdf (retrieved on March 26, 2011).
69. Nationwide Permit (27) Stream and Wetland Restoration Activities (3/18/2002), at http://www.nao.usace.army.mil/technical%20services/Regulatory%20branch/NW_enclosures/NW-27.pdf (retrieved on March 26, 2011).
70. Nationwide Permit (27) Stream and Wetland Restoration Activities (3/18/2002), at http://www.nao.usace.army.mil/technical%20services/Regulatory%20branch/NW_enclosures/NW-27.pdf (retrieved on March 26, 2011).
71. U.S. Environmental Protection Agency National Pollutant Discharge Elimination System (NPDES), at http://cfpub.epa.gov/npdes/stateinfo.cfm (retrieved on March 2, 2011).
72. Conserving America's Wetlands, Implementing the President's Goal. 2005. Council on Environmental Quality, Executive Office of the President. April.
73. Ducks Unlimited Mission Statement, at http://www.ducks.org (retrieved November 1, 2009).
74. Steven Lawrence and Reina Mukai, *Foundation Growth and Giving Estimates*, 2009 Edition, Foundation Center, 2009 at http://www.foundationcenter.org (retrieved November 1, 2009).

About the Author

Tom Biebighauser has established and/or restored over 1,400 emergent, ephemeral, forested, and wet-meadow wetlands on public and private lands in 18 states and two Canadian provinces. He carries a deep and long -standing concern for environmental issues, and thus, finds it rewarding to repair wetlands that have failed, and to assist like-minded land managers who are initiating wetland and stream restoration programs.

Tom enjoys leading workshops where participants learn about wetland restoration by actually constructing wetlands, and has instructed hands-on wetland training in Alabama, Arizona, California, Colorado, Illinois, Kentucky, Maryland, Massachusetts, Minnesota, Missouri, Mississippi, New Mexico, New York, North Carolina, Oregon, Pennsylvania, South Carolina, Tennessee, Virginia, West Virginia, British Columbia, and Ontario. He has also enjoyed helping over 80 teachers build wetlands at their schools for use as outdoor classrooms, and encourages hundreds of students each year to put on waders and venture into wetlands for exploration.

Tom's passion for wetland and stream restoration has resulted in his receiving numerous awards during his career, including recognition for outstanding accomplishments in resource conservation practices from Goodyear Tire and the National Association of Conservation Districts, as well as three National Taking Wing Awards for leadership, community involvement, and outstanding accomplishments. He also was awarded a National Wings Across the Americas Conservation Award for significant contributions to bat management and wetland restoration.

In 2003, Tom wrote and published the book *A Guide to Creating Vernal Ponds,* which he used in his workshops as a training guide. His second book, *Wetland Drainage, Restoration, and Repair*, was well received upon its release in 2007.

Tom and his wife Dee have two married children and two grandchildren. They are both active in their church, where Tom is a Sunday School teacher. When he isn't exploring and photographing wetlands, he can be found making improvements to their farm in Kentucky, which Dee has aptly named *Twisted Arm Farm*.

He invites you to visit the Center for Wetlands and Stream Restoration website: www.wetlandsandstreamrestoration.org, or to e-mail him at tombiebighauser@gmail.com.

INDEX

A

aerial photographs, 26-27
American woodcock, 8, 134
amphibians
  affected by drainage, 11
  affected by vehicles, 12
aquatic plants
  and wetlands, 8-9, 17-18, 67, 79, 122, 135
  control, 50, 75, 127-128, 147
  and drainage, 10, 12, 143, 152
  establishment, 4, 90-91, 94, 107-108, 117, 119, 123, 138
  as wetland indicator, 12,79, 138, 152
archeology. *See* cultural resources

B

backhoe, 36-37, 47, 52, 56, 68, 72, 74-76, 80, 84, 87, 90, 95, 114, 127, 141, 149, 153, 159, 163
bats, 7, 8, 29, 31, 107, 120, 134-135
  Indiana, 3, 70, 135, 159
beaver
  and wetlands, 7, 9, 41, 79, 97, 101, 123, 136, 159
  damage to dams, 43, 47, 75-76, 143, 148, 152
  damage to spillways, 38, 45, 64-65, 77-78, 97, 102-103, 144, 148
bentonite, 154
bogs. *See* wetland types

C

Canada geese, 7, 30-31, 123, 128, 134, 148
carp, 11-12, 75-76, 97, 147
cattails
  control of, 31, 75-76, 123, 126-128, 131
  importance of, 7, 76, 126
  indicator of wetlands, 31, 93, 143, 183
  problem in wetlands, 31, 38, 76, 126, 135, 148
cattle. *See* livestock
changing wetland types, 79
contracting
  considerations, 39, 41, 44, 48, 75, 115, 141-142, 147, 159, 160
  specifications, 54, 118, 165-174
core. *See* groundwater dams
crayfish
  as indicator of drained wetlands, 15
  cause of wetland failure, 40, 48-50, 91, 115, 145, 148, 154
  factor in wetland construction, 51, 55, 73, 80, 82, 103, 170
  value in wetlands, 78, 135, 138
cultural resources, 140-141

D

ditches
  purpose of, 11, 16-18
  locations of, 15-16
  of diversion type, 16
  repairing, 95-96, 99-100, 119
dozer, 46-47, 67
dragonflies, 3, 8, 12, 76, 92, 107-108, 117, 120, 126, 129, 135
drain lines
  as wetland indicator, 12
  locating in ground, 13-14
  of clay tiles, 12-15, 17, 19, 22, 28, 47, 89, 145, 148, 158
  of plastic, 12-14, 19, 51, 78, 81, 114, 145, 147, 159
  of rock and stone, 14-15, 148
  of wood, 14, 19, 22, 51, 78, 81, 148, 159
  outlets, 12, 14, 15, 17-18, 21, 52, 75, 145, 164
drainage districts, 15
drained wetland, identification of, 10-26, 28, 37
drainpipes. *See* water control structures
Ducks Unlimited, Inc., 3
dump truck
  capacities, 74, 103, 110
  use of, 66, 86, 103, 110, 118

E

endangered species, 1, 3, 28-29, 70, 79, 138, 141-143
erosion control, 121, 127
excavator, 46-47, 67
explosives for wetland construction, 87-88

F

fencing, 46, 51, 106, 106, 108, 119, 123, 126, 152
fens. *See* wetland types
fill (construction)
    building on, 22, 47, 74, 93-94, 105, 108
    of wetlands, 4, 7, 10, 12, 16-21, 23, 37, 78, 93, 119, 138, 148, 159
fish
    control of, 48, 75, 97, 147, 152
    habitat improvement, 3, 38, 79, 108, 110, 126, 148, 155, 157-158, 162, 174
    affect on wildlife, 11, 78, 93, 107, 129, 131-132, 135-136, 138, 149
flashboard riser, 76-77, 144
floodplain permits, 139
frogs
    bull, 78, 97, 131
    green, 131
    red-legged, 131-132
    wood, 4-5, 29, 31, 78, 97, 131
funding sources and strategies, 161-162

G

geo-textile, 49, 95-96, 108, 111, 115-118, 147, 163
groundwater dams
    construction, 51-63, 68, 80-81, 90-91 171-173
    equipment required, 36, 46-47, 67, 103
    for stream restoration, 158-159
    for wetland repair, 145-147
    importance of, 49-51, 74-75, 99, 101-102, 105, 163
groundwater wetland construction technique, 35, 79-96

H

head-cuts, 21, 33, 93, 95, 105, 108, 144, 156, 157-159, 163-164
Hickenbottom surface inlet, 13

I

invasive plants, 12, 31, 48, 60, 64, 67, 110, 117, 121-122, 138

K

key to wetland construction techniques, 37

L

lands plowing, 18, 26
level (construction), 32, 43, 44-45, 111-112, 115-117
leveling of land, 12, 16, 19, 47, 84, 138
LiDAR, 25
liner wetland construction technique, 107-119
livestock, 3, 18-19, 68, 78, 87-89, 108, 115, 130, 148-149, 152

M

mined land, 3, 5, 28-29, 103, 130, 135
mole plow, 14
mosquitoes, 107-108, 135
muskrat
    damage to dams, 43, 45, 47, 59-60, 65, 75-76, 92, 97, 102, 123, 135, 143, 148, 151-154
    value to wetland, 7, 12, 127, 129, 135

N

National Environmental Policy Act (NEPA), 137
National Historic Preservation Act. *See* cultural resources
native plant establishment, 122-126
Natural Resources Conservation Service (NRCS), 24, 38-39, 122
non-profit organizations, 161-162

O

oats, 60, 121

P

permits, government, 137-142
ponds, rebuilding, 148-152
posthole digger, 35-37, 79, 88, 91, 123, 153
Prior Conversion (PC) designation, 24, 26, 138
pumps, 15, 38, 69, 76, 79

R

rain delay, avoiding, 48
red-legged frog, 131-132
roads
    construction and wetlands, 11
    ruts, 11-12, 23-24, 30-31, 48, 68, 70-72, 101, 115, 174
rye, 60, 117, 121-122

S

salamanders
    four-toed, 132
    habitat, 7-8, 11, 31, 89, 92, 107-108, 112, 120, 127, 129, 149, 159
    Jefferson, 131
    marbled, 70, 132
    mole, 4
    spotted, 4, 29, 76, 131-132, 138
    streamside, 139
    tiger, 97
school-based wetlands

appearance, 5, 80, 96, 107, 109, 120
construction, 4, 22, 45-47, 96, 99, 107-120, 149, 153
value of, 9, 107, 119
scraper, 47, 78, 85-86, 102-103, 147
Section 401 Permit, 139
Section 404 Permit, 137-138
seeders, 121-122
shorebirds, 3, 7, 31, 38, 75-76, 79, 97, 134, 152, 174
sink holes, 22
snags, 129
soil auger, 35-36, 67, 79, 87-88, 91, 164
soil survey, 30, 40, 138, 159
soil(s)
color in drained wetlands, 22-23
color in wetlands, 7-9, 22-23
texture for wetland construction, 36-37, 40
spikes, landscape, 112-113, 115, 117-118
spillway
and stream characteristics, 90
construction, 38-39, 44-45, 60-61, 64-66, 72, 92, 99, 117, 140, 148, 153, 171, 173-174
importance of, 37, 41, 43, 67, 70-71, 108, 144, 153
springs
draining, 18-19
importance to wetlands, 18
restoration of, 89-91, 130
straw
appearance of, 66, 83, 119, 127
application of, 84, 86, 96, 100-101, 114, 121-122
for erosion control, 60, 64, 68, 70, 87, 121-122, 139-140, 152, 171, 173, 175
streams
moving, 19-21, 155-157
problems with damming, 31, 37-38, 41
restoration of, 155-160
storm water permits, 139-149
storm water wetland development, 91-98, 119
surface inlet, 13, 27, 138, 184
surface water construction technique, 35, 40-78

T

test holes
shallow, 30-31, 35-36, 40, 69, 79, 87-89, 91, 102, 108, 110, 137, 145, 153
deep, 37, 47, 49, 54, 75, 80, 85, 91, 111, 141, 145, 147, 159
tile probe, 36, 93, 110, 115, 164
toads, 8, 11, 87, 89, 92, 107, 114, 119, 120, 128-130, 151
topsoil
for filling wetlands, 21-22, 36-37, 67
value of, 29, 36, 38-39, 48, 66-67, 105, 122, 139
track-steer, 47, 151, 184
tree planting, 123-126
turtles, 3, 7-8, 31, 67, 70, 82, 86, 89, 120, 129, 132,133, 149

U

U.S. Army Corps of Engineers, 137-139
U.S.D.A. Forest Service, 4, 27, 72, 88, 181
U.S. Fish and Wildlife Service, 3, 7, 38, 141-142

V

vehicle damage to wetlands, 11-12, 88-89
vernal pond. *See* wetland types, ephemeral

W

water control structures
installation of, 77, 173, 180
problems, 38, 41, 45, 91-92, 143-144, 148
value of, 37, 75, 78, 97, 102-103, 127, 134, 152-153, 161
water table elevation, determination of, 79-80
wells, 15, 38, 130, 159
wetland
failure, 4
functions of, 3,7
loss from drainage, 3, 7, 23
wetland construction
adding woody debris, 21, 32, 48, 59, 70, 72-74, 80, 86, 93, 100, 113, 115, 127, 129, 151, 158-159
and blocking ditches, 78
and clearing, 48-49, 80-81
and compaction, 68-69
and coring. *See* groundwater dams
and dam location, 43-45
and limestone, 22, 52, 145
and permeable layers, 49-51
and shaping, 41-43
and topsoil, 66-67
and watershed size, 40-41
at schools, 107-120
contracting, 48, 54, 75, 141, 147, 160, 165-178
dam building, 59-64
drainage before, 74-75
floodplain considerations, 45-46, 67-68
groundwater dams. *See* groundwater dams
groundwater technique, 35-36, 79-96
on roads, 70-72
on various topographic locations, 33-35
over construction fill, 22, 47, 74, 93-94, 105-108
site selection, 31-33, 107-108
spillways, 64-66
surface water technique, 35-36
surveying and marking, 41-45, 80-81
using clay liners, 72-73
using synthetic liners, 107-119
wetlands, repairing of, 143-154
Wetland Reserve Program (WRP), 38-39
wetland restoration locations
on fields, 13, 15, 17-19, 23, 27-29, 38, 42

on golf courses, 30
on mined areas, 3, 5, 28, 103, 130, 135
on old roads, 30
on timber harvest areas, 29-30, 93, 98, 128, 131
on transmission line right-of-ways, 30
wetland types, restoration of, 7-9
ephemeral construction, 97-99
emergent construction, 97
fens, 103-104
forested construction, 101-103
peat land construction, 103 -108
shrub construction, 101-102
wet-meadow construction, 99-100
wetlands, value of, 7-9
wheat
application rates, 60, 118
importance for erosion control, 60, 64, 66, 68, 70, 83-84, 86, 87, 93, 100, 114, 117-118, 121, 140, 152
wood ducks, 3, 31, 39, 76, 120, 133-134, 151, 159